Lecture Notes in Computer Science 9881

Commenced Publication in 1973
Founding and Former Series Editors:
Gerhard Goos, Juris Hartmanis, and Jan van Leeuwen

More information about this series at http://www.springer.com/series/7409

Sven Casteleyn · Peter Dolog
Cesare Pautasso (Eds.)

Current Trends
in Web Engineering

ICWE 2016 International Workshops
DUI, TELERISE, SoWeMine, and Liquid Web
Lugano, Switzerland, June 6–9, 2016
Revised Selected Papers

Editors
Sven Casteleyn
GEOTEC Research Group
University Jaime I
Castellón de la Plana
Spain

Peter Dolog
Department of Computer Science
Aalborg University
Aalborg
Denmark

Cesare Pautasso
Faculty of Informatics
University of Lugano
Lugano
Switzerland

ISSN 0302-9743   ISSN 1611-3349   (electronic)
Lecture Notes in Computer Science
ISBN 978-3-319-46962-1   ISBN 978-3-319-46963-8   (eBook)
DOI 10.1007/978-3-319-46963-8

Library of Congress Control Number: 2016953215

LNCS Sublibrary: SL3 – Information Systems and Applications, incl. Internet/Web, and HCI

Printed on acid-free paper

This Springer imprint is published by Springer Nature
The registered company is Springer International Publishing AG
The registered company address is: Gewerbestrasse 11, 6330 Cham, Switzerland

Foreword

The International Conference on Web Engineering (ICWE) aims to promote research and scientific exchange related to Web engineering, and to bring together researchers and practitioners from various disciplines in academia and industry in order to tackle emerging challenges in the engineering of Web applications and associated technologies, as well as to assess the impact of these technologies on society, media, and culture.

This volume collects the papers presented at the workshops co-located with the 16th International Conference on Web Engineering (ICWE 2016), held during June 6–9, 2016, in Lugano, Switzerland. In the tradition of previous ICWE conferences, the workshops complement the main conference, and provide a forum for researchers and practitioners to discuss emerging topics, both within the ICWE community and at the crossroads with other communities. As a result, we accepted six workshops, of which the following four contributed papers to this volume:

- 2nd International Workshop on TEchnical and LEgal aspects of data pRIvacy and SEcurity (TELERISE 2016)
- 2nd International Workshop on Mining the Social Web (SoWeMine 2016)
- 1st International Workshop on Liquid Multi-Device Software for the Web (LiquidWS 2016)
- 5th Workshop on Distributed User Interfaces: Distributing Interactions (DUI 2016)

TELERISE 2016 collected papers discussing legal aspects of the Web, hereby focusing on issues such as data management, security, privacy, copyrights, and intellectual property rights. By reconciling the technical and legal perspectives, TELERISE lived up to the cross-disciplinary spirit of ICWE workshops. SoWeMine 2016 brought together researchers addressing engineering challenges related to social Web mining and associated applications. This workshop too embodies the cross-boundary nature of ICWE workshops, marrying data mining and application engineering disciplines. LiquidWS 2016 addressed the emerging topic of multi-device, decentralized Web applications, in which users seamlessly move from one device to another, and their applications and data seamlessly flows among them. Approaching the topic from a Web engineering perspective, LiquidWS brought together papers tackling architectural and engineering issues, as well as practical example applications. Finally, the DUI 2016 workshop shed light on distributed user interfaces in the multi-device Web. In the fifth edition of the DUI workshop series, the organizers specifically focused on distributed interactions, and succeeded in assembling papers addressing theoretical and practical issues alike.

In addition to the four aforementioned workshops, the ICWE conference also hosted the ICWE2016 Rapid Mashup Challenge (RMC 2016), which traditionally has its own volume published as proceedings, and the 7th International Workshop on Web APIs and RESTful design (WS-REST 2016) which had a working session format with focus on collaboration and discussions, rather than paper presentations. All aforementioned workshops had a rigorous peer-review procedure with only quality papers accepted.

Special thanks are extended to ICWE's sponsors: the Faculty of Informatics at Università della Svizzera italiana, City of Lugano, Google, Nokia, Atomikos, InnoQ, lastminute.comgroup and ISWE, all of whose support made ICWE and the associated workshops possible. We are also grateful to Springer for publishing this workshop volume and for sponsoring travel grants to support student authors. In addition, we thank all the workshop organizers for their excellent work in identifying cutting-edge and cross-disciplinary topics in the rapidly moving field of Web engineering, and organizing inspiring workshops around them. A word of thanks also to the reviewers, for their meticulous work in selecting the best papers to be presented. Last, but not least, we would like to thank the authors who submitted their work to the workshops and all the participants who contributed to the success of these events.

July 2016

Sven Casteleyn
Peter Dolog
Cesare Pautasso

Sponsors

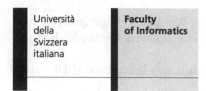

Preface

The preface of this volume collects the prefaces of the proceedings of the individual workshops. The actual workshop papers, grouped by event, can be found in the body of this volume.

2nd International Workshop on TEchnical and LEgal aspects of data pRIvacy and SEcurity (TELERISE 2016)

Organizers. Ilaria Matteucci, Paolo Mori, Marinella Petrocchi, Istituto di Informatica e Telematica – Consiglio Nazionale delle Ricerche (IIT-CNR), Pisa, Italy.

The present volume includes the proceedings of the 2nd International Workshop on TEchnical and LEgal aspects of data pRIvacy and SEcurity (TELERISE 2016), held in conjunction with the 16th International Conference on Web Engineering (ICWE 2016), on June 9 at Università della Svizzera Italiana (USI), Lugano, Switzerland.

TELERISE aims at providing a forum for researchers, engineers, and legal experts, in academia as well as in industry, to foster an exchange of research results, experiences, and products in the area of privacy preserving, secure data management, and engineering on the Web, from a technical and legal perspective. The ultimate goal is to conceive new trends and ideas on designing, implementing, and evaluating solutions for privacy-preserving information sharing, with a view to the cross-relations between ICT and regulatory aspects of data management and engineering. Information sharing on the Web is essential for today's business and societal transactions. Nevertheless, such sharing should not violate the security and privacy requirements either dictated by law to protect data subjects or by internal regulations provided both at the organization and individual level. An effectual, rapid, and unfailing electronic data sharing among different parties, while protecting legitimate rights on these data, is a key issue with several shades. One of the main goals of TELERISE is to carry forward innovative solutions, such as the design and implementation of new software architectures, software components, and software interfaces, able to fill the gap between technical and legal aspects of data privacy and data security management.

This year, TELERISE received a total of ten submissions from 20 authors of eight countries. Each paper was reviewed by at least three Program Committee members and evaluated according to criteria of relevance, originality, soundness, maturity, and quality of presentation. Decisions were based on the review results and five submissions were accepted as regular papers. We have grouped the accepted papers into two main classes according to their topics: "Security and Privacy Aspects," and "Legal Aspects." The keynote speech was given by Benoit Van Asbroeck, partner in Bird&Bird Intellectual Property practice, based in Brussels, and it was titled "Technical and Legal Aspects of Data Privacy." The talk covered the main areas of interest of the workshop. The program was as follows:

- Session 1. Security and Privacy Aspects

  - Harald Gjermundrød, Ioanna Dionysiou, and Kyriakos Costa. "privacy-Tracker: A Privacy-by-Design GDPR-Compliant Framework with Verifiable Data Traceability Controls."
  - Daniel Schougaard, Nicola Dragoni, and Angelo Spognardi. "Evaluation of Professional Cloud Password Management Tools."
  - Neil Ayeb, Francesco Di Cerbo, and Slim Trabelsi. "Enhancing Access Control Trees for Cloud Computing."

- Keynote Session

  - Benoit Van Asbroeck. "Technical and Legal Aspects of Data Privacy."

- Session 2. Legal Aspects

  - Kevin Mcgillivray, Samson Esayas, and Tobias Mahler. "Is a Picture Worth a Thousand Terms? Visualising Contract Terms and Data Protection Requirements for Cloud Computing Users."
  - Francesca Mauro and Debora Stella. "Brief Overview of the Legal Instruments and the Related Limits for Sharing Data While Complying with the EU Data Protection Law."

The second edition of TELERISE was a real success and an inspiration for future workshops on this new and exciting area of research.

We would like to thank the ICWE Workshops Organizing Committee and collaborators for their precious help in handling all the organizational issues related to the workshop. Our next thanks go to the authors of the submitted papers. Special thanks are finally due to the Program Committee members for the high-quality and objective reviews they provided.

July 2016

Ilaria Matteucci
Paolo Mori
Marinella Petrocchi

Program Committee

Benjamin Aziz	University of Portsmouth, UK
Gianpiero Costantino	IIT-CNR, Italy
Vittoria Cozza	IIT-CNR, Italy
Francesco Di Cerbo	SAP Labs, France
Ioanna Dionysiou	University of Nicosia, Cyprus
Carmen Fernandez Gago	University of Malaga, Spain
Sorren Hanvey	Irish Software Research Centre, Limerick, Ireland
Kuan Hon	Queen Mary University, UK
Jens Jensen	STFC, UK
Erisa Karafili	Imperial College London, UK
Mirko Manea	Hewlett Packard Enterprise Italy, Italy

Aaron Massey — Georgia Institute of Technology, USA
Kevin McGillivray — University of Oslo, Norway
Roberto Sanz Requena — Grupo Hospitalario Quiron, Spain
Andrea Saracino — IIT-CNR, Italy
Daniele Sgandurra — Imperial College London, UK
Jatinder Singh — University of Cambridge, UK
Debora Stella — Bird & Bird, Italy
Slim Trabelsi — SAP Labs, France

2nd International Workshop on Mining the Social Web (SoWeMine 2016)

Organizers. Spiros Sirmakessis, Technological Institution of Western Greece, Greece; Maria Rigou, University of Patras, Greece; Evanthia Faliagka, Technological Institution of Western Greece, Greece, Olfa Nasraoui, University of Louisville, USA.

The rapid development of modern information and communication technologies (ICTs) in the past few years and their introduction into people's daily lives have greatly increased the amount of information available at all levels of their social environment.

People have been steadily turning to the social web for social interaction, news and content consumption, networking, and job seeking. As a result, vast amounts of user information are populating the social Web. In light of these developments the social mining workshop aims to study new and innovative techniques and methodologies on social data mining.

Social mining is a relatively new and fast-growing research area, which includes various tasks such as recommendations, personalization, e-recruitment, opinion mining, sentiment analysis, and searching for multimedia data (images, video, etc).

This workshop is aimed at studying (and even going beyond) the state of the art in social Web mining, a field that merges the topics of social network applications and Web mining, which are both major topics of interest for ICWE. The basic scope is to create a forum for professionals and researchers in the fields of personalization, Web search, text mining etc. to discuss the application of their techniques and methodologies in this new and very promising research area.

The workshop tried to encourage a discussion on new emergent issues related to current trends derived from the creation and use of modern Web applications. The following papers were presented:

- Evanthia Faliagka, Maria Rigou, and Spiros Sirmakessis: "Identifying Great Teachers Through Their Online Presence." Teacher evaluation is a very tricky task as there are many criteria, objective and not, that are important in identifying the suitability of a teacher to a specific class. A teacher's background as well his or her education and experience, personality, and even the students of the class are some of the important criteria that take part in the evaluation. In this work, the authors propose a novel approach and a prototype system that extracts a set of objective criteria from the teacher's LinkedIn profile, and infers their personality characteristics using linguistic analysis on their Facebook and Twitter posts.
- Paolo Missier, Alexander Romanovsky, Tudor Miu, Atinder Pal, Michael Daniilakis, Alessandro Garcia, Diego Cedrim, and Leonardo Da Silva: "Tracking Dengue Epidemics Using Twitter Content Classification and Topic Modelling."
  The paper used Twitter for a very interesting topic detection: mosquito-borne diseases. Detecting and preventing outbreaks of mosquito-borne diseases such as dengue and Zika in Brazil and other tropical regions has long been a priority for governments in affected areas. Streaming social media content, such as Twitter, is

increasingly being used for health vigilance applications, such as flu detection. The authors contrast two complementary approaches to detecting Twitter content that are relevant for Dengue outbreak detection, namely, supervised classification and unsupervised clustering using topic modelling.

- Vittoria Cozza, Van Tien Hoang, Marinella Petrocchi, and Angelo Spognardi: "Experimental Measures of News Personalization in Google News." The authors present their work with filter bubbles. Search engines and social media keep trace of profile- and behavioral-based distinct signals of their users, to provide them with personalized and recommended content. The authors focus on the level of Web search personalization, to estimate the risk of trapping the user into these filter bubbles with experimentation carried out on the Google News platform. The aim of the paper is to measure the level of personalization delivered under different contexts: logged users, expected (in SGY sections), and unexpected (in Google News home) personalization.

July 2016

Spiros Sirmakessis
Maria Rigou
Evanthia Faliagka
Olfa Nasraoui
Marinella Petrocchi

Program Committee

Evanthia Faliagka	Technological Educational Institution of Western Greece, Greece
John Garofalakis	University of Patras, Greece
Koutheair Khribi	ALECSO Organization, Tunisia
Maja Pivec	University of Applied Sciences FH Joanneum, Austria
Maria Rigkou	University of Patras, Greece
Muhammet Demirbilek	Suleyman Demirel University, Turkey
Olfa Nasraoui	University of Louisville, USA
Paolo Crippa	Università Politecnica delle Marche, Italy
Spiros Sioutas	Ionian University, Greece
Spiros Sirmakessis	Technological Educational Institution of Western, Greece
Zanifa Omary	The Institute of Finance Management, Tanzania

1st International Workshop on Liquid Multi-Device Software for the Web (LiquidWS 2016)

Organizers. Kari Systä, Tommi Mikkonen, Tampere University of Technology, Finland; Cesare Pautasso, USI Lugano, Switzerland; Antero Taivalsaari, Nokia Technologies, Finland.

The era of standalone computing devices is coming to an end. Device shipment trends indicate that the number of Web-enabled devices other than PCs and smartphones will grow rapidly. In the future, people will commonly use various types of Internet-connected devices in their daily lives. Unlike today, no single device will dominate the user's digital life. In general, the world of computing is rapidly evolving from traditional client-server architectures to decentralized multi-device architectures in which people use various types of Web-enabled client devices, and data are stored simultaneously in numerous devices and cloud-based services. This new era will dramatically raise the expectations for device interoperability, implying significant changes for software architecture as well. Most importantly, a multi-device software architecture should minimize the burden that the users currently have in keeping devices in sync. Ideally, when the users move from one device to another, they should be able to seamlessly continue doing what they were doing previously, e.g., continue playing the same game, watching the same movie, or listening to the same song on the other device. This way the users can take full advantage of all their devices, either using them together at the same time or switching between them at different times.

By "liquid software," we refer to an approach in which applications and data can seamlessly from one device to another, allowing the users to roam freely across all the computing devices that they have. The users of liquid software do not need to worry about data copying, manual synchronization of device settings, application installation, or other burdensome device management tasks. Rather, things should work with minimal effort. From the software development perspective, liquid software should dynamically adapt to the set of devices that are available to run it, as opposed to responsive software, which adapts to different devices, under the assumption that only one device at a time is used to run the application.

The 1st International Workshop on Liquid Multi-Device Software was arranged to present the latest research and discuss the aforementioned topics from the Web engineering point of view. The workshop was held on June 8, 2016, and it was co-located with International Conference in Web Engineering (ICWE 2016) in Lugano, Switzerland. We envision that HTML5 and Web technologies will be used as the basis for a broader, industry-wide multi-device software architecture, enabling seamless usage of applications not only with devices from a certain manufacturer or native ecosystem, but more broadly across the entire industry. HTML5 and Web technologies could serve as the common denominator and technology enabler that would bridge the gaps between currently separate device and computing ecosystems.

After the peer-review process, four papers were selected to be presented at the workshop. The papers covered various aspects of liquid software sharing a focus on user interface design challenges.

The first paper was "XD-Bike: A Cross-Device Repository of Mountain Biking Routes" by Maria Husmann, Linda Di Geronimo, and Moira Norrie from ETH Zrich. The paper presented by Maria Husmann showed how multiple devices can collaboratively provide the users with the needed information. The system used a Web-based framework (XD-MVC) for building MVC cross-device applications. This presentation included a nice demonstration, too.

The second paper was "Multi-Device UI Development for Task-Continuous Cross-Channel Web Applications" by Enes Yigitbas, Thomas Kern, Patrick Urban, and Stefan Sauer from Paderborn University and Wincor Nixdorf. The paper – presented by Enes Yigithas – continued the theme of multi-device user interfaces and described how bank customers can use different devices in different contexts. The researchers were targeting a system in which bank customers are able to flexibly access their banking service – where, when, and how the service suits them best.

The third paper "Liquid Context: Migrating the User's Context Across Devices" by Javier Berrocal, Jose Garcia-Alonso, Carlos Canal, and Juan Manuel Murillo Rodriguez from the University of Extremadura and the University of Malaga extended the discussions to the management of user context. This paper, presented by Javier Berrocal, explained how the user profile and preferences should be taken into account in liquid applications and how the context information should be available wherever the applications migrate.

The fourth paper "Synchronizing Application State Using Virtual DOM Trees" by Jari-Pekka Voutilainen from Gofore Ltd., and Tommi Mikkonen and Kari Systä from Tampere University of Technology described one solution for synchronization of the application state. The paper was presented by Jari-Pekka Voutilainen and it described how a virtual DOM tree can be used to implement state synchronization for liquid applications.

We are grateful to the Program Committee members for their work on the paper review and selection process. We would also like to thank all the authors and workshop participants for the lively discussions.

July 2016   Kari Systä
                                                                   Tommi Mikkonen
                                                                    Cesare Pautasso
                                                                  Antero Taivalsaari

Program Committee

Zoran Budimac	University of Novi Sad, Serbia
Robert Hirschfeld	Hasso Plattner Institut, Potsdam University, Germany
Mirjana Ivanovic	University of Novi Sad, Serbia
Tommi Mikkonen	Tampere University of Technology, Finland
Juan Manuel Murillo Rodriguez	Universidad de Extremadura, Spain
Cesare Pautasso	USI Lugano, Switzerland

Kari Systä   Tampere University of Technology, Finland
Antero Taivalsaari   Nokia Technologies, Finland
Hallvard Trætteberg   Norwegian University of Science and Technology,
                            Trondheim, Norway
Daniele Bonetta   Oracle Labs, USA
Michael Nebeling   Carnegie Mellon University, USA

5th Workshop on Distributed User Interfaces: Distributing Interactions (DUI 2016)

Organizers. María D. Lozano, José A. Gallud, Víctor M.R. Penichet, Ricardo Tesoriero, Computer Systems Department, University of Castilla-La Mancha, Albacete, Spain; Jean Vanderdonck, Catholique Univesity of Louvain, Belgium; Habib M. Fardoun, King AbdulAziz University, Jeddah, Saudi Arabia; Juan Enrique Garrido, Computer Science Research Institute, University of Castilla-La Mancha, Albacete, Spain; Félix Albertos Marco, Computer Systems Department, University of Castilla-La Mancha, Albacete, Spain.

The 5th Workshop on Distributed User Interfaces was focused on distributing interactions. Current technology and ICT models generate configurations in which the same user interface can be offered through different interactions. These new technological ecosystems appear as a result of the existence of many heterogeneous devices and interaction mechanisms. Consequently, new conditions and possibilities arise, which not only affects the distribution of the user interfaces but also the distribution of the user's interactions. Thus, we shift the focus from addressing the distribution of user interfaces to the distribution of the user's interactions, which poses new challenges that need to be explored. In this context, Web engineering appears as a fundamental research field since it helps to develop device-independent Web applications with user interfaces that are capable of being distributed and accessed through different interaction modes. This fact makes Web environments especially interesting within the scope of this workshop. As in the previous workshops in this series, the main goal is to bring together people working on distributed interactions and enable them to share their knowledge in aspects related to new interaction paradigms such as movement-based interaction, speech recognition, gestures, touch and tangible interaction, etc., and the way we can manage them in a distributed setting.

The workshop started with Session 1, which was a somewhat mad session in which each participant introduced himself/herself. This session continued with two research presentations:

- Michael Krug and Martin Gaedke: "AttributeLinking: Exploiting Attributes for Inter-Component Communication." The authors propose exploiting attributes of client-side Web components to provide inter-component communication by external configuration. With the integration of a multi-device supporting MessagingService, components can even be linked across multiple connected devices. This enables the development of distributed user interfaces.
- Juan Enrique Garrido Navarro, Victor M. R. Penichet, and Maria-Dolores Lozano: "Improving Context-Awareness in Healthcare Through Distributed Interactions." This paper describes a significant step forward in the concept of context-awareness with a comprehensive solution: Ubi4Health. The solution enhances context-awareness by adapting the user experience with the appropriate device, interface, and interaction mechanism on the basis of the given context.

Session 2 took place with six presentations:

- Amira Bouabid, Sophie Lepreux, and Christophe Kolski: "Distributed Tabletops: Study Involving Two RFID Tabletops with Generic Tangible Objects." This paper describes a study on an innovative system designed to support remote collaborative games running on tabletops with tangible interaction. In addition, the authors model a set of collaborative styles that are possible between the tabletops users. The goal is to obtain objects that provide remote collaboration among users of interactive tabletops for tangible interaction.
- Félix Albertos Marco, Víctor M.R. Penichet, and Jose A. Gallud: "Distributing Interaction in Responsive Cross-Device Applications." In this work the authors introduce the foundations of a new approach called responsive cross-device applications (RCDA). RCDA applies the idea of responsive Web applications distributing user interactions across the new cross-device ecosystem, taking into account the interactive capacities of devices and users.
- Audrey Sanctorum and Beat Signer: "Towards User-Defined Cross-Device Interaction."
  The authors provide an overview of existing DUI approaches and classify the different solutions. In addition, they propose an approach for user-defined cross-device interaction where users can author their customized user interfaces based on a hypermedia metamodel and the concept of active components.
- Antonio Jesús Fernández-García, Luis Iribarne, Antonio Corral, Javier Criado, and James Z. Wang: "Optimally Storing the User Interaction in Mashup Interfaces Within a Relational Database." Storing the data generated from the interaction performed over the user interface can be challenging. To achieve this goal, in this paper a relational database for storing this interaction information generated on distributed user interfaces is proposed.
- Félix Albertos Marco, Víctor M.R. Penichet, and Jose A. Gallud: "Virtual Spatially Aware Shared Displays." In this work, the authors present a technique for distributing content and devices in shared workspaces using cross-device displays. This technique, referred to as the virtual spatially aware technique, allows the creation of virtual shared displays and the coordination of cross-device interactions. By using this technique, they propose a method for arranging content and devices on virtual displays.
- Sergio Firmenich, Gabriela Bosetti, Gustavo Rossi, and Marco Winckler: "Flexible Distribution of Existing Web Interfaces: An Architecture Involving Developers and End-Users." This paper describes an architecture that allows end-users to collect UI objects into a distributed UIComponent-oriented PIM, accessible from different users' devices. Once in the PIM, different DUI-based behaviors (that may be triggered by the user) are added to the collected UI components as PIM object plug-ins.

The workshop finished with an interesting Session 3, in which the participants collaborated by working together. The objective was to discuss the main ideas and results from the previous sessions, future research lines, and possible collaborations. The organization of the sessions involved all the participants. In particular, during Sessions 1 and 2, the participants listed concepts to be considered in the last session on post-it notes. These concepts were stuck on a board and categorized in Session 3. This activity allowed participants to discuss definitions, links, related and future concepts, etc. The results were an interesting exchange of ideas. Finally, this collaborative work involved the possibility of continuing to collaborate as an initial community related to distributed user interfaces and the topics included in the workshop.

July 2016

María D. Lozano
José A. Gallud
Víctor M.R. Penichet
Ricardo Tesoriero
Jean Vanderdonck
Habib M. Fardoun
Juan Enrique Garrido
Félix Albertos Marco

Program Committee

María D. Lozano	University of Castilla-La Mancha, Spain
José A. Gallud	University of Castilla-La Mancha, Spain
Víctor M.R. Penichet	University of Castilla-La Mancha, Spain
Ricardo Tesoriero	University of Castilla-La Mancha, Spain
Jean Vanderdonck	Université catholique de Louvain, Belgium
Habib M. Fardoun	King AbdulAziz University, Saudi Arabia
Juan Enrique Garrido	University of Castilla-La Mancha, Spain
Félix Albertos Marco	University of Castilla-La Mancha, Spain

Contents

2nd International Workshop on TEchnical and LEgal aspects of data pRIvacy and SEcurity (TELERISE 2016)

privacyTracker: A Privacy-by-Design GDPR-Compliant Framework with Verifiable Data Traceability Controls

Harald Gjermundrød(✉), Ioanna Dionysiou, and Kyriakos Costa

Department of Computer Science, School of Sciences and Engineering,
University of Nicosia, Nicosia, Cyprus
{harald,dionysiou.i}@unic.ac.cy,
kyriakoskosta@gmail.com

Abstract. Breach or lack of online privacy has become almost a commonplace of today's digital age, mainly due to the inability of either enforcing privacy requirements or imposing strict sanctions against violations. The current state of affairs in data privacy is at a turning point for companies operating in EU state members as the enforcement of the General Data Protection Regulation (GDPR) empowers users with control over their personal data, including regulating its disclosure, withdrawing disclosure consent at any given time and tracking their data trail. Compliance with the GDPR is mandatory and it requires significant amendments and/or restructuring of data processing routines undertaken by enterprises. Currently, there is no framework to support the GDPR principles. This paper proposes privacyTracker, a GDPR-compliant framework that supports basic GDPR principles including data traceability and allowing a user to get a cryptographically verifiable snapshot of his/her data trail.

Keywords: User privacy · Data traceability · General Data Protection Regulation (GDPR)

1 Introduction

With the proliferation of digital technologies and the growing trend of digitizing all kinds of records (*e.g.* business, academic, medical, government) concerns over privacy issues are raised not only by organized groups but also by average users of technological solutions, who have a keen interest in the processing and handling procedures of personal data by organizations. According to the 2015 TRUSTe US Consumer Confidence Index [1], 92 % of the respondents worry about their privacy online, revealing as the top cause of concern the companies collecting and sharing personal information with other companies. Consumers want to be informed on how their personal data is used as well as be allowed to stop being contacted by third parties (30 %). Almost half of the respondents stated the need of clear procedures for removing personal information.

© Springer International Publishing AG 2016
S. Casteleyn et al. (Eds.): ICWE 2016 Workshops, LNCS 9881, pp. 3–15, 2016.
DOI: 10.1007/978-3-319-46963-8_1

Privacy, as defined by Westin [2], is the *"claim of individuals, groups, or institutions to determine for themselves when, how, and to what extent information about them is communicated to others"*. Personal data protection is of utmost importance and must be safeguarded, especially online. Usually, online privacy is expressed as privacy policies posted on sites that outline what data is collected, why is collected and how it is used. However, more often than not doubt is cast on their effectiveness. Reasons include, among others, the complexity of the policies themselves that could create more confusion than clarification and the lack of awareness among users with regard to privacy matters. Furthermore, even though the privacy policies are available to the users, there could be a discrepancy between policy statements and their actual implementation. As a consequence, the user is at no position to *verify* that his privacy is properly handled by an organization.

Serious steps should be taken to offer guarantees for user data protection, especially in the light of the new European Council General Data Protection Regulation (GDPR) [3] that was approved in December 2015. Many businesses, most likely, will need to change their data processing practices to conform with the GDPR principles, which empower users not only with the *control* of their own personal data but also with practical *certainty* of their desired access controls. The control extends to include the *right to erasure*, where the user has the right to request erasure of personal data related to him/her under certain conditions. Technical measures must be in place to manage proper data collection and processing, including mapping legal requirements to policies, mapping policies to technical mechanisms, requiring explicit user consent for all collected personal data, updating user personal data to maintain its accuracy, disclosing personal data according to user control preferences, providing personal data traceability upon user request, certifying an enterprise as GDPR-compliant, and honoring the right to erasure, where the user has the right to request erasure of personal data related to him/her under certain conditions. The technical implementation of all GDPR requirements is not trivial, as it requires a complicated framework that maps the legal requirements into technical mechanisms and measures.

As of today, to the best of our knowledge, there is no such framework in place (data protection by design) that complies with the GDPR principles of data collection and processing. Furthermore, there is no compliance checking procedure to oversee the adherence to the regulation policies. Inspired by the GDPR, an ecosystem is proposed in this paper, that supports the collection, trade, and distribution of personal and other consumer data along the lines of the GDPR. At the same time, the ecosystem allows enterprises to create trusted relationships with their consumers based on transparency and verifiable proofs, when required, and remain relevant in the emergent sharing economy. To be more specific, the paper contributions are twofold: presenting the design principles of a GDPR-compliant framework that handles data processing by enterprises and discussing their practicality via the Implementation of privacyTracker, a privacy-by-design GDPR-compliant system.

The remainder of this paper is as follows. Section 2 gives an overview of personal data protection in terms of policies and legislation. Section 3 introduces privacyTracker, a novel framework compliant to GDPR principles and Sect. 4 presents a privacyTracker prototype. Section 5 concludes the paper.

2   Personal Data Protection Overview

The common approach, followed by organizations and companies, to user data privacy is the use of privacy policies. These are usually posted on the organization's main site or are presented to the user, who in turn has to give consent before allowed to proceed with a transaction. There is a plethora of research efforts on privacy policies mostly focusing on (1) formalizing privacy policies that could be analyzed for illegal disclosure and potential conflicts, (2) investigating the effectiveness of privacy policies, (3) privacy policy compliance frameworks and (4) provenance of data [4–8].

The absence of privacy policies or their failure to comply to data protection directives and legislations often lead in violation of user privacy. Additionally, the uncontrolled sharing of information and their aggregation from various sources pose non-negligent threats to user privacy as it yields in constructing user profiles without the user's consent. The examples below demonstrate that indeed privacy policies are no silver bullet in safeguarding one's privacy:

- **Absence of privacy policies:** a recent example comes from an audit of the websites of the 2016 US presidential candidates, conducted by the Electronic Privacy Information Center (EPIC), that found out 4 sites had no stated privacy policy at all [9] and several others did not state their data disclosure practices.
- **Violation of Privacy Regulations:** On February 2015, a report that has been commissioned by the Belgian Data Protection Authority found that Facebook is acting in violation of European law [10]. According to the report, *users are offered no choice whatsoever with regard to the sharing of location data.*
- **Potential Violation of Privacy Regulations:** Security firm AVG can sell search and browser history data to advertisers in order to "make money" from its free antivirus software, a change to its privacy policy has confirmed. The updated policy explained that AVG was allowed to collect "non-personal data", which could then be sold to third parties. The new privacy policy came into effect on 15 October 2015, but AVG explained that the ability to collect search history data had also been included in previous privacy policies, albeit with different wording.

Even in the case where privacy policies are enforced and accurately translated into actual implementation statements that do not compromise the stated privacy, still the user is not aware of his/her personal and other data distribution. There is no practical mechanism that permits the active participation of users in carrying out a formal inquiry on the *whereabouts* of their personal data collected by organizations. This is a serious flaw in the current data privacy frameworks.

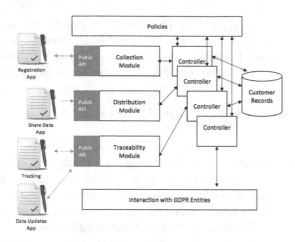

Fig. 1. privacyTracker framework

The current state of lack of accountability when it comes to preserving personal data privacy is about to change as the European Commission General Data Protection Regulation (GDPR), put forward in 2012, attempts to reform the data protection rights across the European Union. An agreement of the proposed regulation was reached on December 2015 and, once it receives formal adoption by the EU parliament and council, its rules will be in effect after 2 years. The GDPR will replace the existing legal framework Directive 95/46/EC and it aims to strengthen citizens' rights to data privacy by giving them control over their personal data.

Any framework that adheres to the GDPR principles must, at a bare minimum, satisfy those data processing requirements (Articles 5(1a), 5(1d), 6(1a), 6(1c), 7(1), 7(3), 12(1), 12(2), 14(1a), 14(1ac), 14a(2g), 15, 16(1), 17(1), 17(2a), 17a(1), 18(2), 19(2)) where the enterprise is obligated to provide undisputed evidence on the handling and sharing of consumer data. This involves addressing the following issues regarding the data in question:

1. be able to accurately set the data collection time and the identity of the collector
2. be able to provide a list of all entities that posses a copy of the original data
3. be able to determine modifications on the data, if any
4. be able to determine the data accuracy and validity, with mechanisms on how to address inaccuracy and invalid data
5. be able to configure the data lifetime, with controls to allow data owners to request data to be erased (right to be forgotten)

Currently, it is nontrivial to get answers to any of the inquiries stated above (except perhaps the first one). Reasons include, among others, the lack of technical solutions, inadequate mandatory legal frameworks that support privacy regarding citizen data and in some cases, lack of interest from the citizen himself on privacy matters. The presented research effort addresses the first obstacle, that of insufficient technical approaches.

3 privacyTracker - A GDPR-Compliant Framework

This section presents the design and implementation details of *privacyTracker*, a privacy-by-design framework that addresses the GDPR data processing requirements. This work follows similar ides to how [11] addressed the involvement of the citizens in an eGovernment setting. Figure 1 depicts the main modules of *privacyTracker*. Details on the main 3 modules are given below (Collection, Distribution, Traceability), along with information on the auxiliary data structure, the Customer Record, which is the core building block of *privacyTracker*. Any framework compliant with the GDPR principles must be policy-driven, thus configurable. This explains the presence of the Policy module that governs the data collection, distribution, and management procedures. Furthermore, provision for interactions with other GDPR entities such as supervisory authorities, data protection officer and the European data protection board could be integrated in the framework.

3.1 Customer Record

The main auxiliary data structure of privacyTracker is the *Customer Record*, a multi-linked list of records that keeps user data encoded in the XML data formate, conforming to the definition of the XML Schema Definition Language (XSD). The advantage of using the open standard self-describing data formate is its portability, thus ease of integration with other applications. The *Customer Record* fields are organized in two sections, the mandatory metadata section and the optional section. The metadata section is comprised of record identification fields, data tractability fields as well as cryptographic controls to ensure data integrity, authenticity, and nonrepudiation. The optional section consists of user public data, user private data that user consent was given for disclosure, data provided by the enterprise itself, to just name a few optional fields. The *Customer Record* metadata fields are defined as follows:

Record Identification

- URI (Unique Resource Identifier) - string concatenation of company name, user email address and auto-generated random identifier. This value is unique within the entire framework, but changes whenever the record is distributed to another entity. Thus, a user may be associated with several URIs.
- User Email Address - could be replaced by a digital signature in the future.
- Genesis Time - timestamp of the initial creation of the record. This value is immutable throughout the framework.
- Creation Time - timestamp of the creation of the record locally. This value is mutable as each company, upon receiving a record, creates a new one locally.
- Expiration Time - record data is considered outdated after this time.

Data Tractability

- Backward-to-Root Reference - A backward reference (link) to the originator entity of the record.

- Backward Reference - A backward reference (link) to the entity that the record was obtained from.
- Forward References - A list of forward references (links) to all entities that this record was disclosed to from its present location.

Cryptographic Controls

- Original Record - A copy of the received signed record.
- Signature - Hash code of the complete record (excluding the original record) signed with the current entity's signing key.

Figure 2 illustrates a record shared among 4 companies, forming a 3-level tree. The root of the tree is *Company A* that created the original record. *Company A* directly shares it with *Company B*, which in turn discloses the record to *Company C* and *Company D*. The bidirectional solid lines between companies represent the forward and backward references while the directed stippled lines represent the backward reference to the root of the tree.

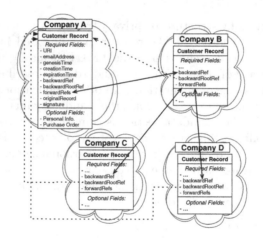

Fig. 2. Customer record tree

Using the example of Fig. 2, the *Customer Record* as it is stored by *Company B* is shown in Listing 1.1. There is a backward root reference to *Company A*, which was the originator of the record as well as a backward reference to the same entity as it is the one that provided the record. Additionally, as *Company B* forwarded the record to both *Company C* and *Company D*, the latter two entities are included in the forward reference list. For brevity reasons, the parent record field is not shown as this is an exact copy of the record stored by *Company A*.

Listing 1.1. Partial Customer Record Document

```
1   <?xml version="1.0" encoding="ISO-8859-1"?>
2   <custRecord rec:URI="www.CompB.com:JohnDoe:20151025_120500" xmlns:rec = "http://
        www.unic.ac.cy/customerRecord">
3       <rec:emailAddress>johnDoe@mail.com</rec:emailAddress>
4       <rec:genesisTime>201510151205</rec:genesisTime>
5       <rec:creationTime>201510251205</rec:creationTime>
6       <rec:expirationTime>201810151205</rec:expirationTime>
7       <rec:bwRef>www.CompA.com:JohnDoe:20151015_120500</rec:bwRef>
8       <rec:bwRootRef>www.CompA.com:JohnDoe:20151015_120500</rec:bwRootRef>
9       <rec:fwRefs>
10          <custRecordList:fwRef>www.CompC.com:JohnDoe:20151028_120500</custRecordList:
               fwRef>
11          <custRecordList:fwRef>www.CompD.com:JohnDoe:20151029_121520</custRecordList:
               fwRef>
12      </rec:fwRefs>
13      <rec:parentRecord>...</rec:parentRecord>
14      <rec:signature>uWta23rFsAEw56Sefgs34 </rec:signature>
15      ...
16  </custRecord>
```

The structure and controls embedded in *Customer Record* allows for utilization of standard generic tree operations for tree traversal and construction of data trails. Furthermore, record removal as well as update operations are possible via the forward references kept in the record. Needless to say, in a real deployment, deeper and broader trees would be constructed per customer record.

3.2   Collection Module

The *Collection* module is the data collection point of *privacyTracker*. Customizable registration applications interact with this module via its public API. There is no automated way to examine whether or not the collected data is lawful and adhering to legal state/country processing laws. Thus, for maintainability purposes, low coupling is strived between the registration application and the Collection module. That implies user consent is obtained via the customized registration application and the data communicated to the Collection module is flagged as disclosed or non-disclosed, based on the user preferences. Each new registration results in the creation of a new customer record. Any optional fields that are outcomes of further data processing or user-company transactions are assessed for legality by the controller module. Similarly to the data collection legality issue, it is beyond the scope of this research effort to automate the legality of data processing. However, the provision of the placeholder could accommodate a future automated routine as a plugin.

3.3   Distribution Module

The Distribution module manages requests to share customer data, either in coarse-grained manner or fine-grained manner. Similar to the previous module, transfer data requests are submitted via a custom application that interfaces with the module API. The requestor could form customized queries on preferred data transfers or use predefined queries. The receiving entity evaluates the request, which leads to 3 possible course of actions: reject, accept as received or partially accept by filtering out records and/or record fields that are not to be disclosed. The latter option gives control to the owner of the data records to decide their further disclosure, even when the data owners gave consent for its disclosure.

As a record gets distributed and handled by many entities, undisputed verifiable guarantees must be provided regarding the record integrity. Any record modifications should be attributed to the entity that made the changes. This is achieved via cryptographic techniques, and to be more specific by digitally signing the hash of the customer record. A company could potentially modify a record in order to incorporate additional data and/or change existing ones and share the new version with others rather than forwarding the version it obtained. Prior to distribution, the original record is embedded in the new record as one of the metadata cryptographic control fields and the hash of the new record is generated, signed, and inserted as the second metadata cryptographic control field (that was signed by the company that disclosed the record). The embedded cryptographic controls provide for nonrepudiation as a user would be able to gather all available versions of his/her record (via the traversal algorithm described later on) and a company could not deny the existence of record versions originated from it. Note that companies receiving a record from the same source must possess the same original record, regardless of any further changes that they may do on the record.

3.4   Traceability Module

A core element of any proposed GDPR-compliant framework is the ability to trace data from its original source to various destinations. Data traceability requires the collaboration of all enterprises and has two components: tracking and tracing. Tracking is the capability to record the path of data as it gets shared with other companies other than the source company that collected the data. Tracing is the capability to identify the origin of data and needless to say tracing will only be successful with properly implemented tracking. Data traceability is the building block to support a variety of GDPR requirements, including the right to erasure and providing the original source of the data.

The proposed framework supports data traceability by utilizing two references of the customer record metadata. When the organization (source) is about to share the record with another organization (target), the source company places a Forward Reference in the record metadata that points to the location that the target company will use to store the record. Similarly, the target organization upon record transfer, inserts a Backward Reference into the metadata of the new record that it creates locally, which points back to the record of the source company. This process is repeated whenever the record is shared. As a result, an *implicit* tree is created (see Fig. 2), with the root node being the originator of the data.

In addition to the forward and backward references there is also a Backward-to-Root Reference in all the records. The reason for maintaining the backward-to-root reference is for recovery reasons in case there should be a link breakage somewhere along the record trail. Link breakage is interpreted as company unavailability or unreachability during contact attempts. A variety of reasons could cause this situation, including out-of-business and legal issues. Using the backward-to-root reference, the unavailable link is located and the repair mechanism is initiated. With the backup backward-to-root reference then this breakage could be located and a repair could be initiated.

It is important to note that whereas a user has the legal right to traverse the record tree, from root to the branches, companies should only be allowed to traverse one level up or one level down the tree (parent node or child nodes) to preserve user privacy. This is a default setting in the *privacyTracker* and access controls are in place to implement this restriction (it could be lifted if deemed necessary).

Below, details are given on constructing the data trail for a specific user, repairing unreachable link references, and addressing the right-to-erasure; all operations are mapped into generic tree operations.

Construction of Data Trail. The construction of a data trail is a standard generic tree traversal problem. Algorithm 1 depicts the steps to traverse the *Customer Record* implicit tree in bottom-up approach, starting from any tree node (i.e. any company that holds the record) towards the root of the tree (i.e. the original creator of the record). The end result is a path from any node to the root.

Algorithm 1. Traverse Customer Record Algorithm

```
 1: function TRAVERSE(CustRecordURL url, EmailAdr adr)
 2:     CustRecordURI parentURI ← null
 3:     CustRecord parentRecord ← null
 4:     CustRecordURI currentURI ← GETCUSTRECURI(url, adr)
 5:     CustRecord currentRecord ← GETCUSTREC(currentURI, adr)

       ▷ Loop backward until reach root
 6:     while (currentRecord != null) do
 7:         SHOWRECORD(record)
 8:         parentURI ← GETPARENTURI(currentRecord)
 9:         parentRecord ← GETCUSTREC(parentURI, adr)

       ▷ Test for broken link
10:         if (parentRecord = null and parentURI != null) then
11:             REPAIRTREE(currentRecord, adr)
12:         else
           ▷ Check for tampering with record
13:             if (VERIFYREC(currentRecord, parentRecord) = false) then
14:                 REPORTVIOLATION(currentRecord, parentRecord)
15:             end if
16:         end if
17:         currentRecord ← parentRecord
18:     end while
19:     SHOWSUMMARY(void)
20: end function
```

Suppose a customer receives an unsolicited request from *Company D*. Traversing the path from *Company D* to the root, the customer could discover who originally collected the data and how the original record was propagated from company to company to end up in *Company D*. Along this path one should be able to determine who disclosed the record unlawfully. The algorithm requires two input variables: the *url* of the company that sent the solicitation and the user's email address. The company gets a customer record request and returns the customer record URI which the user can use for the request to return the whole customer record (see lines 4–5). The backward tracing starts as a repetition process (see lines 6–18). The parent record is first obtained. In case the parent record is *null*, but the parentURI is not *null*, then a breakage in the tree has taken place. In this scenario, the tree repair algorithm is initiated (details below). If there is no breakage in the tree, then a validation check is done (see line 13) to test the integrity of the record contents compared to the parent record contents. If such a modification took place, a violation is reported to the user. It is outside the scope of the framework, for the time being, to investigate how violations are addressed. The last line in the repetition process (see line 17) is used to move one level up in the tree towards the root.

A user has the right to obtain from an organization all the recipients to whom his/her data have been disclosed. A similar algorithm could be used to search the tree top-down (using breadth or depth first search) in the opposite direction. Suppose that the user desires to view all recipients of his/her data starting from a specific company. In this case, a forward searching algorithm will be used (not included here) with the end result being a tree.

Recovering from Unavailable Link References. The repair algorithm works like a standard *remove node* from a double linked list. Suppose that the parent node of the current node is unavailable, thus the link references must be updated so as the current node will have backward reference to its grandparent node. This entails using the backward-to-root reference to perform a forward search to locate the grandparent of the current node and readjust the link references. The assumption is that no other nodes in the tree are unavailable. In the unlikely scenario where 2 nodes on the data trail are unavailable, two different approaches could be deployed to reestablish connectivity in the tree, with different tradeoffs.

Right-to-Erasure. The right-to-erasure requires erasure of user data from all its recipients. With the current data structure, this is easily implemented by constructing a tree for the user data starting from the root to all its leaves, and proceed with deleting all versions for the particular user along all tree paths.

4   A privacyTracker Prototype

A prototype was built along the principles of *privacyTracker* as a proof-of-concept regarding the feasibility of the proposed approach. The prototype is a web-application consisting of three modules, built on top of a WAMP (Windows, Apache, MySQL, PHP) server. Additional technologies used are JavaScript, CSS, XML, HTML 5, MD5 hashing algorithm, and OpenSSL. The experimental setting consisted of 6 companies.

Collection Module: The collection module, depicted in Fig. 3, allows user registration. There are 3 ways that user data could be communicated to the *privacyTracker*. First, directly using the prototype's registration form. In this case, data validation is supported (e.g. address format in different countries) via regular expressions, followed by insertion into the backend MySQL database. Second, having customized registration modules using the provided API to populate the database. Third, through the distribution module (presented next), where traded data is merged with the local company data. It could be the case that multiple entries exist for a single user. The database consists of 3 tables and is normalized to support this. PHP scripts generate the tables in the database, hence there is no need of manual management of the database.

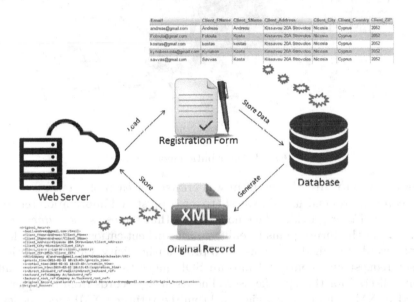

Fig. 3. Registration module

Distribution Module: The distribution module is responsible for the sharing/selling/trading customer information and it is divided into 3 submodules. The first submodule accepts requests for data transfers, which are translated into SQL queries. The prototype supports a web view where the user manually specifies the information to be traded and the receiver entity. The selection of data to be shared is illustrated in Fig. 4. The second submodule encodes the result of the SQL query into an XML document, digitally signed by the current enterprise. The signed document is transferred to the receiving organization using an SSL channel. Once the document is received, the sending company proceeds with updating the forward references of the successfully transmitted records. The last submodule is executed by the receiving company that, upon verification of the XML signed document, converts it to SQL statements that populate the recipient database with the new data. In addition, the backward reference is created to point to the sending company. The received XML document is also saved into the permanent log directory. In the case that the receiving company already has information about a user (identified by the email address), the user-specific records are merged. In the unlikely scenario that the exact same record already existed, the company keeps its own original copy. This could happen if a lattice is created; for example *company A* sells a record to *company B* and *company C*; then *company D* buys the same record from both *company B* and *company C*.

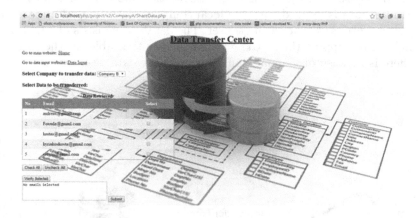

Fig. 4. Distribution module

Traceability Module: A web form was created for each of the six companies that accommodates end-users' requests to query on stored information related to them. The end-user provides the email address that serves as the *authentication* token. It is in the future plans to enhance the authentication process with one-time passwords (emailed to the user) to prove authenticity. Once authenticated the user request gets converted into an SQL query that returns all the information collected for this specific user. The resulting records from the query are encoded in XML and digitally signed. From the returned XML document, the end-user can use the forward and/or backward references to build a trace tree. It is envisioned that user apps will be created to automatically build the complete trace tree from any starting point. The *privacyTracker* framework provides the appropriate APIs and hooks for the development of such apps.

5   Conclusion

To the best of our knowledge, there is no practical mechanism that determines accurately the disclosure of data collected by organizations. There are privacy policies that vaguely specify the handling and processing of data, however the consumer is not informed neither about the identity of the third-party entities that have access to his/her data nor the actual data that is accessible by them. This paper presented the *privacyTracker* framework, a novel approach that empowers consumers with appropriate controls to trace the disclosure of data as collected by companies and assess the integrity of this multi-handled

data. This is accomplished by constructing a tree-like data structure of all entities that received the digital record, while maintaining references that allow
traversal of the tree from any node, both in top-down manner and bottom-up
manner. A prototype was developed based on the *privacyTracker* principles as
a proof-of-concept of the viability of the proposed principles.

Acknowledgment. The authors would like to thank the BeWiser consortium (funded
under EU FP7, Grant No: 319907) for fruitful discussions on citizen security and privacy
issues.

References

1. TRUSTe: 2015 truste us consumer confidence index (2015). https://www.
   truste.com/resources/privacy-research/us-consumer-confidence-index-2015/.
   Accessed 25 Sept 2015
2. Westin, A.: Privacy and Freedom. Atheneum, New York (1967)
3. Parliament, E.: Regulation of the European Parliament and of the Council on the
   Protection of Individuals with regard to the Processing of Personal Data and on
   the Free Movement of Such Data (General Data Protection Regulation). Technical
   report (2015)
4. Karjoth, G., Schunter, M., Waidner, M.: Platform for enterprise privacy practices:
   privacy-enabled management of customer data. In: Dingledine, R., Syverson, P.F.
   (eds.) PET 2002. LNCS, vol. 2482, pp. 69 84. Springer, Heidelberg (2003)
5. Kalloniatis, C., Mouratidis, H., Vassilis, M., Islam, S., Gritzalis, S., Kavakli, E.:
   Towards the design of secure and privacy-oriented information systems in the cloud:
   Identifying the major concepts. Comput. Stand. Interfaces **36**(4), 759–775 (2014),
   Security in Information Systems: Advances and new Challenges
6. Barth, A., Datta, A., Mitchell, J.C., Nissenbaum, H.: Privacy and contextual
   integrity: framework and applications. In: 2006 IEEE Symposium on Security and
   Privacy Security and Privacy, pp. 184–198 (2006)
7. Bertino, E., Ghinita, G., Kantarcioglu, M., Nguyen, D., Park, J., Sandhu, R.,
   Sultana, S., Thuraisingham, B., Xu, S.: A roadmap for privacy-enhanced secure
   data provenance. J. Intell. Inf. Syst. **43**(3), 481–501 (2014)
8. Mont, M.C., Pearson, S., Bramhall, P.: Towards accountable management of identity and privacy: sticky policies and enforceable tracing services. In: 2003 Proceedings of 14th International Workshop on Database and Expert Systems Applications, pp. 377–382 (2003)
9. Epic: Electronic privacy information center survey: 74% of presidential candidate's
   websites fail on privacy. https://epic.org/2015/09/survey-74-of-presidential-cand.
   html. Accessed 25 Sept 2015 (2015)
10. Alsenoy, B.V., Verdoodt, V., Heyman, R., Ausloos, J.,Wauters, E.: From social
    media service to advertising network: a critical analysis of facebook's revised poli-
    cies and terms. Technicalreport, Interdisciplinary Centre for Law and ICT/Centre
    for Intellectual Property Rights of KU Leuven and the department of Studies on
    Media of the Vrije Universiteit Brussel (2015)
11. Gjermundrød, H., Dionysiou, I.: A conceptual framework for configurable privacy-
    awareness in a citizen-centric egovernment. Electron. Gov. **11**(4), 258–282 (2015)

Evaluation of Professional Cloud Password Management Tools

Daniel Schougaard[1], Nicola Dragoni[1,2(✉)], and Angelo Spognardi[1]

[1] DTU Compute, Technical University of Denmark, Lyngby, Denmark
ndra@dtu.dk
[2] Centre for Applied Autonomous Sensor Systems, Örebro University,
Örebro, Sweden

Abstract. Strong passwords have been preached since decades. However, lot of the regular users of IT systems resort to simple and repetitive passwords, especially nowadays in the "service era". To help alleviate this problem, a new class of software grew popular: password managers. Since their introduction, password managers have slowly been migrating into the cloud. In this paper we review and analyze current professional password managers in the cloud. We discuss several functional and non-functional requirements to evaluate existing solutions and we sum up their strengths and weaknesses. The main conclusion is that a silver bullet solution is not available yet and that this type of tools still deserve a significant research effort from the privacy and security community.

1 Introduction

For many years, IT professionals have preached the importance of strong passwords. Many publications exist, describing exactly what defines a strong password and user habits [1]. The general consensus is that it needs *at least* both upper- and lower-case letters, digits and preferably also symbols *(#, _, etc.)*. Additionally, it should not be a word, or a word where an L is replaced by a 1. And of course it has to be at least 8 characters long. More importantly, the user is not supposed to use the same password for more than one service. With all of these rules for strong passwords, it comes as no surprise that many low-security-educated users of IT services resort to simple and repetitive passwords.

To help alleviate this problem, a new class of software grew popular: Password managers. Those are simple tools, usually protected by a single master password, able to generate and store in a secure manner, distinct and hardly-to-guess passwords in place of the user herself. A lot of the IT professionals took these tools to their heart, despite their inherent —very often hidden— flaws.

As with many other contexts in modern society, the users crave convenience. In particular, tools storing an encrypted file with all the password locally, was no longer sufficient, as the majority of users began to use multiple devices and needed to have passwords available in all of them. Hence, the password managers slowly migrated into the cloud. This also saved the users from the hassle of managing their passwords, themselves: the users unload some of the "responsibilities" onto third parties and their data are kept for them, available at all times, from any device.

S. Casteleyn et al. (Eds.): ICWE 2016 Workshops, LNCS 9881, pp. 16–28, 2016.
DOI: 10.1007/978-3-319-46963-8_2

While the cloud *does* come with its benefits, especially convenience, it has its own drawbacks as well, primarily *trust*. When uploading data into the cloud, the user is effectively trusting the service provider. She is trusting that the provider is completely honest about the inner working of its service, mainly regarding what it can and can not access. Users are trusting the providers when they say that they do not share their information to third parties. Unfortunately, sometimes this trust is betrayed, mainly when service providers experience technical incidents. In the context of cloud password managers, for example, it is well known the involving LastPass company in 2015[1]. As many IT professionals had feared, the online password manager had a breach. Panic arose and LastPass almost forced their users to change their passwords.

However, even if trust is a general issue with the cloud, in the case of password managers it is particularly critical, as the user trusts a service to store confidential information that give access to, potentially, all the other services the user everyday accesses. Thus, it is ultimately important to have a detailed knowledge and a objective security assessment of the password manager services available in the cloud.

Contribution and Outline of the Paper. The main contribution of this paper is a comparative and critical security analysis of the different alternatives available for the user, with the final aim to understand if a suitable manager already exists or if (as it is) further efforts are required to provide adequate protection to users' passwords. In particular, in this paper

- we consider and discuss functional and non-functional requirements for password manager services in the cloud;
- we survey and perform a usability and security assessment of 14 typologies of professional password manager tools available in the cloud;
- we compare the results of the assessment and focus on the main weaknesses

We think that the final outcome of our analysis will raise the awareness of the less-security-aware users and will call the IT community for a higher effort to face the password management in the cloud. We want to stress that the paper is focused on available professional password manager tools, while purely academic approaches are left as future work.

The rest of the paper is organized as follows: next section contains the analysis of the functional and non-functional requirements a password manager service in the cloud should guarantee. Section 3 is focused on the description of the password manager services considered for this paper, while Sect. 4 contains the comparison of the obtained results. Section 5 concludes the paper with some final future directions.

2   Functional and Non-functional Requirements

In this section we report and briefly describe the most desirable requirements a cloud password manager service should have. We distinguish between functional

---

[1] http://krebsonsecurity.com/2015/06/password-manager-lastpass-warns-of-breach/.

and non-functional requirements [20]. The former define the expected functioning of the system, namely what the system is expected to do, while the latter refer to qualities of the system, including performance, usability, reliability and so on. In the next subsection we identified 17 functional requirements, as desirable features of the system.

2.1   Functional Requirements

The first mandatory functionality a cloud password manager solution should have is the accessibility: namely, it should be possible to have access to the passwords from several physical devices. We say that it has to have a *distributed password database*. Another desired functionality is that it should support *multiple* —individual— users. It should also be possible to distinguish between *administrators* and regular users. In order to better restrict outside access, the admin will have to create a new user. This can be done either with the admin actually setting up the user, or an invite to registration.

It should be possible to organize passwords in a *structured way* and in multiple levels, customizable by the individual users, for the best user experience. For convenience, it should also be possible to selectively *share passwords*, according to the user needs.

The desirable solution should be *platform agnostic*, and should not be limited to one specific server software. In particular, the user should be able to choose what type of underlying storage/*database*, he or she prefers to use. This would also make possible to run it on low powered devices.

No password —or any other sensitive data— should *ever be present unencrypted* anywhere else, than a local device. This ensures that even if another part of the solution is somehow compromised, data is not revealed on that device.

The users should be able to audit access to their personal data including, but not limited to, *retrieving* passwords, adding/*changing* passwords, and *deleting* passwords. This should be done leveraging a *logging system*, able at least to record detailed access time and the remote host. This ensures that a user can detect if, when and from where unauthorized accesses have occurred.

Access to the system should be protected by the users *master password*, and it should be *possible to change* it. Enabling and using a *two-factor authentication* mechanism should be a possible option. Finally, to protect the availability of the system, we would require that the client-side of the system should *automatically restart* after a hardware reboot.

2.2   Non-functional Requirements

Considering non-functional requirements, we selected 7 desirable properties of password manager services. Firstly, we would require that there is the option to store the passwords where the user has *control over*. This would make the system more flexible, since it would open the way for a password manager in a private cloud.

In order to promote further development, allowing for use of various open source frameworks and libraries, the solution should be *open source* and licensed with an appropriate license (MIT for instance). The solution should be scalable, namely able to store at least *million of password* entries, spread across all users. The *encryption used for storing* the passwords should be of industry standard, and should be viable for at least 5 years. The same goes for the *encryption used for communication*. For maximum security, the solution should only accept and use *TLS version 1.2* connections, with a limited cipher suite.

Finally, for the best user experience, all the interaction with the user interface should be realized with a *latency never exceeding* 500*ms*. Any longer, and the user will grow tired of using the software, because of its sluggish feel.

3   Tools

In this section we briefly introduce 14 different available password manager tools, detailing the most relevant features and postponing in Sect. 4 a more thorough analysis. We considered only real systems already usable to final users, as listed in Table 1. In the last part of this section, we also report a concise survey of proposals coming from the literature and not available as usable tools.

Table 1. Password managers considered in the analysis

1. In-Browser built-in	6. Zoho Vault	11. SimpleVault
2. LastPass (and similar)	7. TeamPasswordManager	12. RoboForm
3. KeePass (and similar)	8. Passwordstate	13. Vaultier
4. Rattic	9. Simple Safe	14. TeamPass
5. Encryptr	10. PassWork	

In the following sections we briefly describe each of the considered solutions, with also a critical eye towards the user experience and the usability: if the solution is not user friendly, the users will not use it and then it is effectively worthless.

1. In-Browser Password Managers. The most used password managers are probably the ones built-in into the various browsers. This is a feature most major browsers have adopted: Chrome, FireFox, Edge (new name of Internet Explorer), Safari and Opera. Almost all of the most recent versions of the mentioned browsers can sync their passwords between different devices, but this requires to upload the passwords to one of the corporations' Web sites. Additionally, built-in password managers have one big limitation: they only work within web sites accessed through that *specific* type of browser, i.e. only in Chrome browser. Passwords for other applications (like email clients, development suites and so on) cannot be easily retrieved.

In [27] it is presented an analysis of the storage formats for the different browsers' password managers. While their results are for probably outdated versions (for example the analyzed version of Chrome was $v.21.0$, while at the time of writing, the current newest version is $v.47.0$), their primary concern is the encryption methods used by the web browsers to store the passwords. At the time of their analysis, only Firefox and Opera were supporting a master password to enable the access to the stored passwords.

2. LastPass, and Similar Solutions. LastPass[2], PassPack[3], DashLane[4], and many others are smartphone apps coupled with plug-in browser enabling the user to access the passwords from several devices. We refer only to LastPass as a representative of this group, it being the most well-known.

LastPass uses 256-bit AES encryption for the communications and applies PBKDF2, as the hashing function, in order to make it difficult to crack stored data. Both encryption and decryption are performed client side [10], as to avoid transferring the actual password, unencrypted, to their servers. Encryption and decryption are done using the master password, which is never actually sent to their servers. Finally, as is to be expected, all connections to LastPass' servers, are TLS 1.2 encrypted.

Regarding the usability, LastPass allows the user to organize passwords in folders, creating the tree-like structure. For devices without a browser supporting plug-ins, LastPass offers a so-called bookmarklet [9]. A bookmarklet is a bookmark, which essentially contains JavaScript code, in order to add previously unobtainable features, in a browser. While this on the surface seems like a nifty feature, work in [12] discusses an attack on LastPass, exploiting the users bookmarklet, to gain access to virtually all of the users stored credentials. Finally, it is work mentioning that there has been a recent leak from LastPass [25], that leads to even more users to look suspicious of their services.

3. KeePass, and Similar Solutions. KeePass[5] gained fame after the LastPass data breach. Differently to this latter, KeePass allows the user to store the passwords in a local file. While there exists a plethora of tools similar to KeePass, it will be used as a representative of this group.

Version 2.x of KeePass uses AES-256 encryption, but it can also apply additional algorithms through plug-ins [8]. This enables users to tailor the encryption security, to their own requirements. KeePass features a tree-like structure, in order to completely organize passwords and also has a fully customizable password generator, where the user can also choose the character sets.

KeePass lacks of usability, since it does not support password distribution. Since KeePass works on a local file, it would only inherently work on a *single* device.

---

2 https://lastpass.com/.
3 https://www.passpack.com/.
4 https://www.dashlane.com/.
5 https://keepass.org.

Should one wish to distribute it, another tool has to be involved to save the file in the cloud. Additionally, there is the lack of cross-platform compatibility, since KeePass only supports Windows.

4. Rattic. Rattic [7] is a self-hosted password manager, in the so-called private cloud. Rattic can be considered a password management database, with a special focus on managing passwords for a team [7]. Since Rattic *is* meant for teams it has multi-user support and makes the distinction between admin and regular users. It organizes passwords and users in groups, for easy access control, where a group is a collection of users which can access the same passwords. Additionally it supports tags for their passwords, allowing for even further organization, for their users allowing quick access to similar passwords, from across different groups. However, the fact that Rattic is team-oriented, the user cannot simply create "private" passwords, but it needs to manually create a group with a single user. Rattic also provides a password generator and makes possible to download passwords in the KeePass format, making it available for later offline use.

Regarding the technical aspects, Rattic surprisingly does *not* encrypt passwords stored in the database and highly recommends storing the database on an encrypted drive, to ensure database protection. Clearly, this does mean that a system administrator can access *all* passwords, should he or she have the encryption key for the drive. As a positive note, Rattic is developed in Python, using the Django framework and tested on the Apache server.

5. Encryptr. Bordering between the type of LastPass and Rattic, Encryptr [3] relies on the Crypton [6] backend [4]. Crypton is an application framework and backend service to develop applications, providing the required primitives for cryptography. Encryptr can host the passwords on a third party cloud service (namely SpiderOak[6]), but makes also possible to run a dedicated Crypton backend, like a the private cloud. However, this requires a high level of technical skills, including editing source files [5], apply patches, compile and fine setting. This severely affects the usability of the solution. Moreover, the user interface is *very* minimalist and sleek, while passwords are stored in one unique, single list.

Despite its complexity, the Crypton backend stands for its zero-knowledge security [19]: according to the authors, it is impossible to obtain the unencrypted data on their servers, without actually getting hold of the users' private encryption keys. The Crypton backend is open source and uses AES-256.

6. Zoho Vault. Zoho Vault [15] relies on the storage within proprietary cloud and aims at enterprise customers, providing interesting features, such as LDAP integration. Vault organizes passwords in so called "chambers" and each password can be added to one or more chambers. While this approach sounds a valid alternative to the classic tree-style organization, it does not add any real benefit

---

[6] http://spideroak.com

to the final user. Zoho Vault uses a combination of RSA and AES. To enable sharing, a common AES encryption key is retrieved using RSA keypairs [16].

7. Team Password Manager. Team Password Manager [26] is a tool that is fairly similar to Rattic, being enterprise-oriented and organizing passwords in a tree-like hierarchy of "projects" and sub-projects. It can be privately-hosted, and is platform independent, requiring Apache, PHP and MySQL.

Team Password Manager uses AES-256 and bcrypt as its key derivation function, to slow down brute force attacks against password hashes. It also supports Google's Authenticator, for two-factor authentication. Being enterprise-oriented, it is very complex and can keep away low-security-wise regular users.

8. Passwordstate. Passwordstate from ClickStudios [17] is another enterprise-oriented solution. It has Active Directory support, and other built-in options, as High Availability, a fully customizable password generator and several options for two-factor authentication. Passwordstate is self-hostable, but requires a Windows platform and the IIS server (it is not platform agnostic). Passwordstate has a very intuitive UI and it is able to give the most important information, the most screen-space. The tree-like structure that creates an excellent organizational options for the users. However, its enterprise-oriented engine presents the user with many technical options, reducing its usability for a regular user.

Passwordstate encrypts all passwords using AES-256 [18] and protects the backend services (web service and database) employing code obfuscation of the pages and encryption of the data.

9. Simple Safe. Simple Safe [14] is another self-hosted team password manager solution. The user can choose to organize passwords into "groups", that are accessed through a menu, which is auto hidden. Unfortunately, changing between groups takes a long time, in the order of seconds. This causes a bad experience for the user: while generally the developers of Simple Safe have been very generous with the animations, this causes a heavy latency that adds to the overall sluggish feel of the system. Considering the possibility to organize the passwords, it is worth citing as the user can customize the fields in the different groups. For instance, they allow the user to store SSH keys, in dedicated field type, making possible to separate passwords that require keys/certs, from traditional username/password logins.

The technical details about Simple Safe are very few. The only available information regarding security or encryption is the following quote: *"Simple Safe utilizes a 256 bit encryption method. Each password has a unique private salt along with a master salt stored separately to the database. Only encrypted passwords are stored within the database."* [13].

10. PassWork. PassWork [21] is yet another enterprise-oriented solution, available as both a remote and a self-hosted, both of which comes with a price tag. It organizes passwords in "groups" and sub-groups with an associate list of users currently allowed to access that passwords. Interestingly, it offers the option of setting permissions to users, namely when adding a user it is possible to give different permissions like "Full Access", "Edit", "Read" and so on.

Being a closed source solution, PassWork's developers do not describe and document their solution adequately. Something as simple as figuring out which platform it supports, prior to purchasing is not possible. Additionally, they simply state of using "256-bit passwords" and RSA keypairs for sharing passwords.

11. SimpleVault. SimpleVault [2] encrypts each individual password with a different "passphrase". As the name suggests, SimpleVault is much more a proof of concept, with an extremely minimalist interface: adding a password is done in a page having two fields completely not described. It provides a powerful password generator, with increasingly "rare" symbols. Retrieving a password, requires the user to type in the passphrase. The choice of this per-password passphrase, will undoubtedly only result in the user using the same password over and over again. However, this choice *does* allow for the system to be used by multiple users, each just using their own master passphrase for all of their passwords.

The security of SimpleVault is again described with few details. The authors state that at some point or other, the encrypted data are eventually stored in clear text on the remote machine. Moreover, they state that when adding a new entry, the password is sent without encryption to the server, which in turn encrypts it, only relying on the HTTPS. Additionally, SimpleVault does *not* use a database system, but stores all contents in a .txt file.

12. RoboForm. RoboForm [24] is somewhat a hybrid between KeePass and LastPass: it locally stores the password and is available to the user as a web browser extension. Additionally, it offers "RoboForm Everywhere" cloud synchronization of the encrypted file. This does however, entail trusting them with safe-keeping the encrypted file. Using RoboForm on a single device, renders it as a tool very similar to KeePass, albeit with "better" browser integration.

13. Vaultier. Vaultier [22] is self-hosted, has a imperceptible latency for any UI actions, and has a clean design of the front page. Passwords are organized in vaults, where cards are placed and, inside of these cards are the passwords. However, the UI showing inside a single vault, only shows six cards at a time. Similarly, at most two passwords can be displayed on screen, at a time. It also has a password generator, with limited customizable options.

Vaultier is open source, since it has a the git repository's history publicly shown. However, it seems like the project is in a idle state, since the last update was done on the 17th of April, 2015 [23].

14. TeamPass. TeamPass [11] ensures platform agnosticism due to being implemented in PHP, and only requiring an Apache server and MySQL server. It features a series of APIs to permit an access to TeamPass Items from a third party application. However, it only relies on HTTPS to transmit the passwords that are, indeed, sent using the GET HTTP method. Additionally, based on the description of the APIs, it appears clear as the password is decrypted server-side.

4   Analysis

While the previous section provided a high level description of the considered solutions, here we detail how each of them deal with the requirements introduced in Sect. 2, providing a unified picture. The analysis is shown in Fig. 1.

In order to better investigate some of the requirements, we provide a closer look to some of the most desirable functional requirements. Figure 2 distinguishes between solutions that are able to be run in the private cloud, and those who are not. Focusing only on those which actually can be hosted in a private cloud, Fig. 3 shows their default organization with respect to the passwords. While all of the systems support multiple users, they all do not consider personal passwords.

This is primarily a result of nearly all of these being marketed towards enterprises and team-oriented development. However, for almost all of the solutions, it is possible to solve this shortcoming, creating a new personal group for each individual user, in which personal passwords can be stored. Unfortunately, this must be considered as a workaround, with the only exception of PassWork, which creates this group automatically.

Looking at the more technical aspects of the assessment, in Fig. 4 we report the agnosticism of the compared solutions. Most of the password manager service are able to run on multiple platforms, being implemented in PHP. Hence, they only require an Apache (or nginx) Web server to run. Fortunately, both of these have versions for all the major operating systems. Considering the database agnosticism, instead, there is only Rattic that actually offers different options. However, even in this case, the developers warn that different options are done at the user's risk, since all the alternatives have to be considered experimental.

As our analysis shows, *no* tool covers all of the requirements. Although some tools might come close, they still lack too much to be considered a viable option.

5   Conclusion

Cloud password managers are becoming increasingly popular, helping users to protect their services with stronger passwords, relieving them from the extra effort required. The number of possible professional alternatives is high, as this paper shows, but also a perfect solution is still missing. We have considered 17 functional requirements and 7 non-functional requirements as the desirable features a cloud password manager should possess. Consequently, we have surveyed 14 professional password manager solutions and critically analyzed against the considered requirements. We focused on professional tools only, as we aimed at getting a clear picture of available implemented cloud password managers. What emerges is that a silver bullet application is still lacking, as no professional password manager fully satisfies the set of functional and non-functional requirements. Consequently, research work has still to be done to propose a

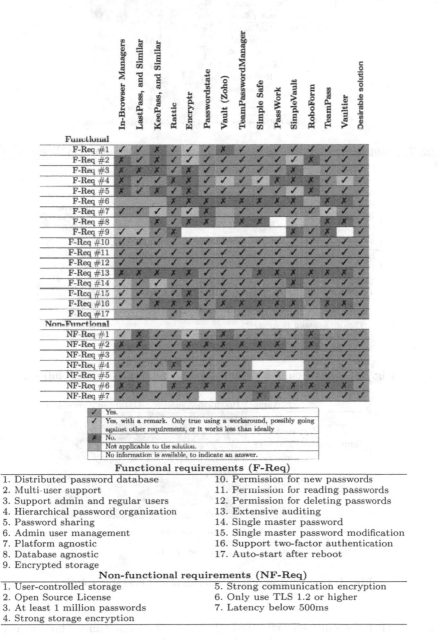

Fig. 1. Analysis of requirements and available professional password manager

Name	Web Based	Self-Hostable
LastPass	Yes	No
KeePass	No	N/A
Rattic	Yes	Yes
Encryptr	No	Yes [a]
Passwordstate	Yes	Yes
Vault *(Zoho)*	Yes	No
TeamPasswordManager	Yes	Yes
Simple Safe	Yes	Yes
Passwork	Yes	Yes
SimpleVault	Yes	Yes
RoboForm	No	N/A
TeamPass	Yes	Yes
Vaultier	Yes	Yes

[a]Requires editing source files and running own Crypton backend.

Fig. 2. Comparison of access method and ownership model, of the available solutions.

Name	Supports Multiple Users	Personal Passwords	Password Sharing
Rattic	Yes	Yes[a]	Yes
Passwordstate	Yes	Yes[a]	Yes
TeamPasswordManager	Yes	Yes[a]	Yes
Simple Safe	Yes	Yes[a]	Yes
Passwork	Yes	Yes	Yes
SimpleVault	Yes[b]	Yes[b]	Yes[c]
TeamPass	Yes	Yes[a]	Yes
Vaultier	Yes	Yes[a]	Yes

[a]Through password grouping.
[b]By each user, using their own unique passphrase for encrypting their personal passwords.
[c]Through a commonly used master passphrase, acting as a workaround.

Fig. 3. Comparison of password ownership in self-hostable solutions.

Name	Platform Agnostic	Database Agnostic
Rattic	Yes	Yes[a]
Passwordstate	No	No
TeamPasswordManager	Yes	No
Simple Safe	Yes	No
Passwork	N/A[b]	N/A[b]
SimpleVault	Yes	No[c]
TeamPass	Yes	No
Vaultier	Yes	No

[a]Databases other than MySQL receives less testing.
[b]No information available to determine.
[c]Doesn't use a database. Stores passwords in a .txt file.

Fig. 4. Comparison of agnosticism of self-hostable solutions.

solution that is secure, effective and efficient at the same time. Thus, we call the security community for a higher effort to face the password management in the cloud, providing adequate protection to users' passwords.

References

1. Amico, M.D., Michiardi, P., Roudier, Y.: Password strength: an empirical analysis. In: 2010 Proceedings of INFOCOM, pp. 1–9. IEEE, March 2010
2. Brugger, R.: Simplevault - password manager. http://simplevault.sourceforge.net/. Accessed 09 Dec 2016
3. Crypton: Encryptr - powered by crypton. https://crypton.io. Accessed 05 Dec 2016
4. Devgeeks: About encryptr - powered by crypton. https://encryptr.org/#about. Accessed 05 Dec 2016
5. Devgeeks: Add server software for self-hosting. issue #156 spideroak/encryptr. https://github.com/SpiderOak/Encryptr/issues/156. Accessed 05 Dec 2016
6. Devgeeks: Crypton - build private applications. https://encryptr.org. Accessed 05 Dec 2016
7. Hall, D.: RatticDB. http://www.ratticdb.org. Accessed 05 Dec 2016
8. KeePass: Keepass. http://keepass.info/help/base/security.html. Accessed 05 Dec 2016
9. LastPass: Bookmarklets — user manual. https://helpdesk.lastpass.com/bookmarklets/. Accessed 10 Dec 2016
10. LastPass: How it works — lastpass. https://lastpass.com/how-it-works/. Accessed 10 Dec 2016
11. Laumaill, N.: Teampass. http://teampass.net/. Accessed 07 Jan 2016
12. Li, Z., He, W., Akhawe, D., Song, D.: The emperors new password manager: security analysis of web-based password managers. Technical report, University of California, Berkely (2014)
13. Ltd., H.P.: Frequently asked questions. https://www.simplesafe.net/faqs/. Accessed 08 Dec 2016
14. Ltd., H.P.: Team password management — simplesafe - self-hosted password sharing. https://www.simplesafe.net/. Accessed 08 Dec 2016
15. Ltd., Z.C.P.: Password manager, password management software - Zoho vault. https://www.zoho.com/vault/. Accessed 05 Dec 2016
16. Ltd., Z.C.P.: Secure sharing of secrets. https://www.zoho.com/vault/secure-sharing-of-secrets.html. Accessed 05 Dec 2016
17. Passwordstate: Enterprise password management with passwordstate. http://www.clickstudios.com.au/. Accessed 08 Dec 2016
18. Passwordstate: Secure code, secure data. http://www.clickstudios.com.au/about/secure-code-data.html. Accessed 08 Dec 2016
19. Pedersen, C., Dahl, D.: Crypton: Zero-knowledge application framework. Technical report, SpiderOak. https://crypton.io/crypton.pdf. Accessed 01 Sep 2016
20. Pressman, R.: Software Engineering: A Practitioner's Approach. McGraw-Hill higher education, New York (2005)
21. Primepix, Konfeta: Team password manager. collaboration and password sharing, API, hostable version. https://passwork.me/. Accessed 05 Dec 2015
22. RightClick: Collaborative password manager & file storage. https://www.vaultier.org. Accessed 17 Dec 2015
23. RightClick: rclick / vaultier — gitlab. http://git.rclick.cz/rclick/vaultier. Accessed 17 Dec 2015
24. Siber Systems, I.: World's best password manager. http://www.roboform.com/. Accessed 05 Dec 2015
25. Siegrist, J.: Lastpass security notice. https://blog.lastpass.com/2015/06/lastpass-security-notice.html. Accessed 05 Dec 2015

26. TeamPasswordManager: Secure sharing of secrets. http://teampasswordmanager. com/. Accessed 05 Dec 2015
27. Zhao, R., Yue, C.: All your browser-saved passwords could belong to us: a security analysis and a cloud-based new design. In: Proceedings of the 3rd ACM Conference on Data and Application Security and Privacy (CODASPY 2013). ACM (2013)

Enhancing Access Control Trees for Cloud Computing

Neil Ayeb, Francesco Di Cerbo$^{(\boxtimes)}$, and Slim Trabelsi

Security Research, SAP Labs France,
805, Av. du Docteur Maurice Donat, 06250 Mougins, France
francesco.di.cerbo@sap.com

Abstract. In their different facets and flavours, cloud services are known for their performance and scalability in the number of users and resources. Cloud computing therefore needs security mechanisms that have the same characteristics. The Access Control Tree (ACT) is an authorization mechanism proposed for cloud services due to its performances and scalability in the number of resources and users. After an initial set-up phase, the ACT permits to simplify the evaluation of an authorization request to a simple visit to the tree structure. Our contribution extends ACT towards instance-based access control models by allowing the expression and evaluation of conditions in access control decisions. We evaluated our contribution against an Open Source authorization mechanism to evaluate its performance and suitability to production settings. Early results seem encouraging with this respect.

Keywords: Access control · Data structures · Cloud

1 Introduction

In all its different facets and variations, cloud computing offers a common set of characteristics like elastic computing and resource pooling [4] in order to serve its tenants and their end-users with high performances. To achieve this objective, cloud service providers commit significant resources, for example in terms of processing and storing. Compelled by normative requirements and service level agreements, cloud services must be secured and protected, as well as any data they host; leaks or abuses may result in severe losses for cloud providers, in terms of liability, fines and reputation on the market.

These simple observations lead us to consider the importance of security aspects in the design, implementation and operation of cloud services. It is possible to derive some basic requirements for security controls in cloud computing, among them:

– **R1** cloud security mechanisms must cope with high volumes of requests and with their quick evaluations, and
– **R2** cloud security mechanisms must be able to cope with high amount of users and resources.

© Springer International Publishing AG 2016
S. Casteleyn et al. (Eds.): ICWE 2016 Workshops, LNCS 9881, pp. 29–38, 2016.
DOI: 10.1007/978-3-319-46963-8_3

We claim that these requirements are particularly important for cloud Access Control (AC) mechanisms: they regulate access to cloud resources and operations, by making authorization decisions depending on parameters like the callee's identity, the requested resource or service and the attempted operation. These parameters and the indications on how to analyse them, compose an AC rule.

These rules can be aggregated in two different forms, through AC lists or policies (from now on ACL). The former approach consists of lists of rules that in essence provide an answer to the question, whether a given user could or not access to a given resource. This solution may result limiting, as potentially many combinations of cloud-hosted resources and end-users for each tenant must be explicitly mentioned in ACL, resulting in complex ACL authoring and maintenance. On the contrary, AC policies can capture rules that specify conditions/actions and offer a powerful, yet conceptual, tool of expressiveness and are particularly useful for complex systems [5]. However, in this case it is required a more sophisticated rule processing, performed through a specific AC engine, in charge of rules interpretation. Ultimately, it must be able to make a reasoning process on the given rules and make a decision: permission or denial. Intuitively, a drawback of expressing sophisticated rules is the need for a reasoning process that could be time consuming. In fact, this has been pointed out in [8,9], namely for resource access in cloud computing, where performance could be seriously affected.

In [9], the authors proposed a new access control mechanism to make fast authorization decisions, based on an high-speed caching tree. The Access Control Tree (ACT) is designed in order to simplify the AC decision making the process to a visit to the ACT. Our contribution consists of an extension to this concept to support a form of instance-based AC, by introducing conditions expressed in AC rules as elements of the ACT. This extension permits to represent more sophisticated policies as part of the decision tree, thus easing the adoption of ACT in more complex scenarios. To illustrate our contribution, Sect. 2 presents a number of findings from previous research initiatives to support the need for cloud-specific AC mechanisms, as well as two brief introductions to XACML and to the ACT. Section 3 presents our extension to ACT while Sect. 4 the results of our performance tests. Lastly, Sect. 5 concludes the paper.

2   State of the Art

2.1   Access Control Mechanisms for Cloud Computing

Many popular cloud services use AC solutions that are not specifically designed for the cloud and for its requirements, especially for **R1** and **R2**. This observation is shared by a number of authors (for instance, [7]). This results in a serious gap that can affect at least the configuration and operation of cloud computing solutions. For this reason, Younis et al. [10] identify performance and scalability as the first requirement for modern cloud-ready AC mechanisms.

Well known cloud services like Amazon S3 or Microsoft Windows Azure Storage, still offer very simple ACL-based solutions, essentially allowing access only to known users or publicly available, without giving the possibility to selectively grant access to principals outside of their domain.

The adoption of caching solutions for cloud mechanisms has been proposed in the past. Reeja [6] propose the usage of two policy decision points (PDP), one for new access requests and another one for access requests that were already processed by the primary PDP. Harnik et al. [3] propose a capability-based system that allows the integration of existing AC solutions thus leading to hybrid architectures for AC systems. Such hybrid systems combine the benefits of capability-based models with other commonly used mechanisms such as ACLs or RBAC. On the other hand this study points out the weakness of ACL used by many major cloud providers like Amazon and Microsoft.

2.2 XACML

XACML [2] is an open standard for the specification of access control policies defined by OASIS.

The standard defines a reference architecture for the implementation of the authorization functionality in an application. Such architecture is composed by a number of components. Among them, we find the Policy Enforcement Point (PEP), a part of the application that intercepts and regulates the access to protected resources or functionalities. The PEP permits an access request to take place if authorized by another component, the Policy Decision Point (PDP); the latter is responsible for making the decisions on the basis of a set of security policies and on the parameters of the request transmitted by the PEP (e.g. the target resource, the desired operation, the entity requesting for it). Such decision-making process considers attributes of the user, of the resource, of the system environment. These attributes are used for evaluating conditions stated in security policies expressed using the XACML policy language; the latter is an XML-based standard designed to support ABAC (Attribute Based Access Control). Policies are stored, maintained and made available to the PDP though another architectural component, the Policy Administration Point (PAP), and are evaluated with attributes coming from different sources (e.g. an LDAP directory containing user's attributes) by means of the Policy Information Point (PIP).

In the XACML policy specification, the policy element can be a policy set, a policy or a rule. A policy set is composed of multiple policies, while a policy consists of a number of rules. Each policy has a Target, that defines with which attributes the policy is applicable (e.g., resource, subject, action). Moreover, the policy contains an 'effect' element that determines if the set of identified attributes are related to a Permit or a Deny. It is also possible to define conditions that will be applied prior returning the decision taken by the engine. The XACML standard (with a particular attention in version 3.0) defines 'obligation' handling. An Obligation is an action that must be executed by the PEP when it receives the decision (e.g., logging the access history for a resource). Figure 1 presents an elaboration of the architecture proposed in [9] for the deployment of XACML AC mechanisms in the cloud. In this setting, the access to cloud resources, be them services, virtual machines, storage or any other, is mediated by the XACML engine that can make decisions considering aspects like multi-tenancy and service-level agreements.

2.3 Access Control Tree

The ACT was introduced by Trabelsi et al. in [9]. The ACT aggregates different policies and their rules in a tree entity in order to make authorization decisions with high performances, through the application of hashing techniques on the tree for efficient data search functions. ACT can be used in an XACML architecture as in Fig. 1 for optimizing the decision making process. For mapping XACML rules into tree elements, Trabelsi et al. adapted a model proposed by Gabillon et al. [1] and adapted it to the XACML policy schema. In their model, only accessible data objects are part of the tree. If the access to an object is denied, it will not appear. This permits to simplify the decision making process and to reduce the number of tree elements. Due to its construction, the tree structure is called 'Permit Tree' and it is indicated in the following with 'P'.

The ACT is composed of four different levels. The first contains the list of authorized subjects (or users, or roles, etc.) declared in the XACML policy repository. The ANY subject ID is used for objects that are accessible to all users with no restrictions. The second level represents the different actions or operations that can be executed on the data. If the list of actions is undefined, there is also an element Any (Action). The third level represents the different types for data objects. Subsequently, the fourth layer contains a list of accessible data object IDs. The layer order (in the example subject, action and resource) can change according to system requirements. For example the first level can be the ID of the object. In that case the selection is made on the object to be accessed, for which one gets the list of authorized users.

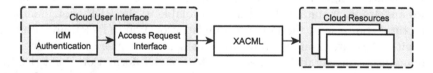

Fig. 1. Architecture for XACML mechanisms in the cloud, adapted from [9]

3 An Extension to ACT

The ACT as presented in [9] consists of a methodology to organise a set of AC directives in the form of a tree structure, in such a way that the evaluation of AC requests boils down to a search navigation in the tree.

The ACT is created by means of an insertion algorithm and queried via a request algorithm. The insertion algorithm allows to construct the tree, structured in four levels: subjects, actions, resource types, resources. The request permits to consider the elements of an AC request and to use them as parameters to search the AC tree. If a result is found, the request can be permitted (in a Permit tree, denied in a Deny tree).

We observed that the ACT achieves the objective of meeting scalability in the number of users and resources together with significant performances. We are focussing this contribution on instance-based Access Control, i.e., methods that permit the expression of fine-grained directives for each individual resource that is under control.

Our proposal has the objective to extend this result further by supporting policies with more sophisticated rules. Therefore, our contribution consists of adding a fifth level to the tree, in order to express the conditions that have to be evaluated at the moment of the decision making process. Multiple conditions may be part of this last layer. A diagram representing the new ACT is depicted in Fig. 2. This new level requires a modification to the ACT insertion and request algorithms, that are used respectively to populate and to query the Permit Tree. A detailed explanation of the algorithms is available in [9].

The new insertion algorithm (that takes advantage of Procedure InsCond) is rendered in Algorithm 1. It starts by considering a set of AC rules in the form $< subject, action, resource >$. If a rule does not include one or more of the tuple constituents, the missing elements are replaced by the special value ÄNY·∴. The algorithm parses each rule and populates the tree by creating a new branch (if necessary) for each touple elements in the respective level.

Our extension considers, for each rule being analysed, whether the rule contains any conditions and in case, they are added to the fifth level of the Permit Tree if they are not already inserted.

Conversely, the request algorithm was modified in order to return the condition(s) that have to be evaluated before issuing an access permission. Algorithm 2 describes this extension.

Procedure InsCond(rule, location)

> **input** : policy rule as R, an ACT node as *Location*
> **foreach** *condition C in R* **do**
> > **if** *C is not in Location* **then**
> > > Add *C* to *Location*

---

4   Evaluation and Preliminary Results

We evaluated our proposal against a standard Open Source AC solution, Balana[1], a well-known XACML engine. We compared the performances of our extended

---

[1] https://github.com/wso2/balana.

ACT and Balana by evaluating a set of up to 3600 policies, generated as the set of all combinations for 60 subjects and 60 resources, performing 7200 requests equally distributed for permit and deny. The experiments have been repeated more than 800 times with different policy settings (minimal parameters were 20

Algorithm 1. Insertion algorithm for ACT

---

input : <R1,R2,...,Rn>as P, *Resource*, current tree as T
Data: *handledRules*={} set of rules of type <subject,action>
foreach *rule R in P* **do**

  if *R is of type <subject,action,permit> OR <subject,any decision>* **then**
    if *T.subject not exist* **then**
      Add subject in *T.subjects*
      Copy actions from *T.any* into *T.subject*

  if *R is of type <subject, action, decision>* **then**
    if *decision == permit* **then**
      Add *Resource* in *T.subject.action*
      InsCond(R, T.subject.action.Resource)
    Add *<subject, action>*in *handledRules*

  else if *R is of type <subject, any, decision>* **then**
    actionList = T.decision.subject.actions
    foreach *action A in actionList* **do**
      if *<subject, A>not in hanledRules* **then**
        if *decision == permit* **then**
          Add *Resource* in *T.subject.A*
          InsCond(R, T.subject.A.Resource)
        Add *<subject, A>*in *handledRules*

  else if *R is of type <any, action,decision>* **then**
    subjectList = T.decision.subjects
    foreach *subject S in subjectList* **do**
      if *<S, action>not in handledRules* **then**
        if *decision == permit* **then**
          Add *Resource* in *T.S.actions*
          InsCond(R, T.S.actions.Resource)
        Add *<S, Actions>*in *handledRules*

  else if *R is of type <any, any, decision>* **then**
    subjectList = T.decision.subjects
    foreach *subject S in subjectList* **do**
      actionList = T.decision.S.actions
      foreach *action A in actionList* **do**
        if *<S, A>not in handledRules* **then**
          if *decision == permit* **then**
            Add *Resource* in *T.S.A*
            InsCond(R, T.S.A.Resource)
          Add *<S, A>*in *handledRules*

Algorithm 2. Request Algorithm for ACT

input : <*subject, action*>as R, *Resource*, current tree as *PT*
Data: *handledRules*={} set of rules of type <subject,action>
if *R.subject not exist in PT* **then**
  Return set of data objects ids resulting form *PT.any.action* and their conditions
else
  Return set of Data object ids resulting form *PT.subject.action* and their conditions

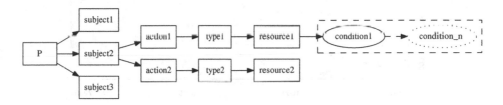

Fig. 2. The extended ACT

users, 20 resources and 800 requests) to confirm the soundness of the results. The policies were generated from the same policy template that contains a rule with a condition requiring for its evaluation an interaction with the PIP to retrieve the value of an attribute. Both ACT and Balana were extended in order to share exactly the same source code for the PIP attribute retrieval operation from a database. The ACT tree structure has been stored in memory.

To simulate a realistic usage in the cloud of the two platforms, we ran our experiments by starting multiple execution threads which request access to resources. For completeness, we also considered the case with a single requestor thread. The number of requests is twice as high as the number of resource/ user combination. This is due to the fact that the requests are intended to cover all the possible cases (i.e., Permit and Deny cases, with and without DB access).

The experimental setting used for the evaluation consisted of an Intel(r) Core(tm) i7 4790 CPU (4 Haswell Cores, 3.75 GHz), 16 GB of System Memory, a 240 GB PCI-EX SSD.

4.1 Performance Tests and Analysis

The first experiment we conducted, depicted in Fig. 3, consisted of measuring the total average time on several executions needed by ACT and Balana to evade 800, 1800, 3200 and 7200 requests when they are issued by single execution thread. The testing scenario of single-threaded sequential query on both of the Balana engine and the ACT shows that the latter always outperforms the former: the difference between ACT and Balana grows from the initial of 5x for 800 requests, to 2 orders of magnitude for 7200 requests. This can be explained by the time

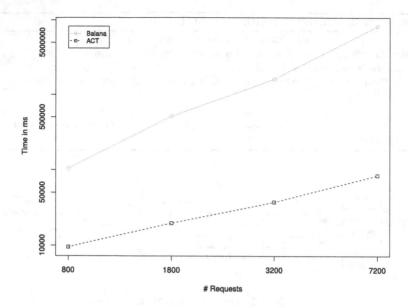

Fig. 3. Comparison of experiments using ACT and Balana

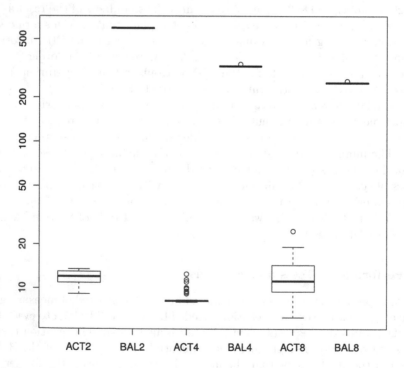

Fig. 4. Execution time in ms for single request using ACT and Balana with 2, 4, and 8 threads.

needed for Balana to scan all the available policies in the repository and trying to match between the request attributes and the policies attributes.

We then conducted a number of experiments in a multi-threaded setting. We present in Fig. 4 the time in ms required for evaluating a single request (out of 7200) using a different number of threads (2, 4 and 8). We calculated these values by collecting the execution times for more than 800 experiments. In all cases, the ACT outperforms Balana, even though the latter scales better in the number of processing threads. On the other hand, the ACT performs slightly more inconsistently in the 8-thread scenario, resulting in slower execution time and a more significant variance as shown by the box plot. Considering that the database used to evaluate the rule condition was deployed on the same execution machine, the fast execution time per single request of ACT as well as the total number of (virtual) cores available on the experimentation server (8), we may conclude that the system resources were almost fully saturated. In the case of Balana, given the higher time necessary for each execution, this effect cannot be observed. The variance of the execution times in Balana experiments is in fact quite limited.

5   Conclusion

Cloud computing needs cloud-designed, performing, efficient and scalable AC solutions. Our proposal extends a cloud AC mechanism based on the XACML standard, the Access Control Tree in order to support instance-based decisions, extending the decision tree with rule conditions. We described our contribution and we evaluated its impact in comparison with an Open Source XACML solution. The results are encouraging and seem aligned with the performance and scalability requirements that we aimed at. As future work we aim at performing a more extensive evaluation and at enhancing further the ACT support for XACML features. Particularly interesting challenges are represented by the evaluation of the impact of rule combining algorithms on the ACT creation process, as well as the possibility to evaluate the impact of multiple conditions (possibily organised in parallel branches) on the performances of the request algorithm.

Acknowledgements. This work was partly supported by EU-funded (FP7/2007–2013) project CoCo Cloud [grant no. 610853].

References

1. Gabillon, A., Munier, M., Bascou, J.-J., Gallon, L., Bruno, E.: An access control model for tree data structures. In: Chan, A.H., Gligor, V.D. (eds.) ISC 2002. LNCS, vol. 2433, p. 117. Springer, Heidelberg (2002)
2. Godik, S., Anderson, A., Parducci, B., Humenn, P., Vajjhala, S.: OASIS eXtensible access control 2 markup language (XACML) 3. Technical report, OASIS (2002)
3. Harnik, D., Kolodner, E.K., Ronen, S., Satran, J., Shulman-Peleg, A., Tal, S.: Secure access mechanism for cloud storage. Scalable Comput. Pract. Exp. **12**(3), 317–336 (2011)

4. Mell. P.M., Grance. T.: SP 800-145. the NIST Definition of Cloud Computing. Technical Report. NIST, Gaithersburg, MD, United States (2011)
5. Popa, L., Yu, M., Ko, S.Y., Ratnasamy, S., Stoica, I.: Cloudpolice: taking access control out of the network. In: Proceedings of the 9th ACM SIGCOMM Workshop on Hot Topics in Networks, p. 7. ACM (2010)
6. Reeja, S.: Role based access control mechanism in cloud computing using cooperative secondary authorization recycling method. Int. J. Emerg. Technol. Adv. Eng. 2(10), 25–34 (2012)
7. Shiftehfar, R., Mechitov, K., Agha, G.: Towards a flexible fine-grained access control system for modern cloud applications. In: IEEE CLOUD, pp. 966–967. IEEE (2014)
8. Tang, Z., Wei, J., Sallam, A., Li, K., Li, R.: A new RBAC based access control model for cloud computing. In: Li, R., Cao, J., Bourgeois, J. (eds.) GPC 2012. LNCS, vol. 7296, pp. 279–288. Springer, Heidelberg (2012)
9. Trabelsi, S., Ecuyer, A., Alvarez, P.C.Y., Di Cerbo, F.: Optimizing access control performance for the cloud. In: Proceedings of CLOSER 2014, pp. 551–558 (2014)
10. Younis, Y.A., Kifayat, K., Merabti, M.: An access control model for cloud computing. J. Inf. Secur. Appl. 19(1), 45–60 (2014)

Is a Picture Worth a Thousand Terms? Visualising Contract Terms and Data Protection Requirements for Cloud Computing Users

Samson Esayas, Tobias Mahler, and Kevin McGillivray[✉]

Norwegian Research Center for Computers and Law, University of Oslo, Oslo, Norway
{s.y.esayas,tobias.mahler,kevin.mcgillivray}@jus.uio.no

Abstract. The following article evaluates two models for providing purchasers of online digital content, including cloud computing services, with visual notice of contract terms and data collection practises. Visualisation of contract terms and privacy policies has the potential to provide cloud consumers with an improved means of understanding the contract terms they are accepting when entering into an agreement with a Cloud Service Provider (CSP). The following paper examines two concrete proposals or models for the visualisation of contract terms and privacy practises as compliance tools in the European context. The article focuses primarily on consumer and data protection law. Although the visualisation models are not currently binding or legally required, they start an important conversation on how such terms can be more effectively conveyed.

Keywords: Visualisation of law · Contract terms · Consumer protection law · Data protection law · Cloud computing

1 Introduction

Cloud computing is being promoted as the future of IT consumption [1]. Promises of cost savings, worldwide availability, and state-of-the-art technologies are enticing to many users [2]. In addition to businesses, consumers are also adopting cloud computing and making changes in the way they consume IT. Despite the many upsides, cloud computing continues to contain challenges for consumers. Many of these challenges revolve around uncertainty regarding privacy, security, access, and portability.

The current barriers to increasing user adoption of cloud computing extend beyond measures that are purely technical or purely legal, including also informational barriers. These barriers have proved significant in Europe, where only 21 % of the population aged 16–74 using the Internet reported using cloud storage [3]. Although this number could be interpreted as encouraging for a relatively young technology, 44 % of users that were aware of cloud computing did not use the services. Of those 'cloud-aware' users not adopting the services, many cited concerns over privacy and the reliability of CSPs as their main reasons for not using cloud computing [3].

© Springer International Publishing AG 2016
S. Casteleyn et al. (Eds.): ICWE 2016 Workshops, LNCS 9881, pp. 39–56, 2016.
DOI: 10.1007/978-3-319-46963-8_4

Many of the risks described above and the protections and obligations that consumers have are designated in the contracts consumers enter into with CSPs [4]. However, given the complexity of the agreements, a general asymmetry of information exists. It is not that consumers necessarily lack information. Reading the quantity of contract terms, privacy policies, and terms of use the average consumer is presented with on a yearly basis is in practise insurmountable [5]. Given the extreme quantity of information, the quality of that information and the legal requirements to provide such information are not always clear. It is not difficult to imagine a situation where a consumer is presented with too much information and suffers from so-called 'information overload'. As a result, contracts, and privacy policies are seldom read.

Although various consumer protection mechanisms require that a substantial amount of information must be provided to consumers regarding their rights under the agreement, many providers of digital content fail to make available the information required by law. Even when such information is provided, it can be difficult for the consumer to find and understand. These rights, if understood and appreciated by consumers, could serve to alleviate some of the concerns consumers have with adopting cloud computing. Additionally, it may provide the would-be reader with a greater ability to comprehend the legal risks inherent in the text [6].

At least one tool—the use of graphical language represented in legal icons—is being actively pursued as a means of providing consumers in the contracting and data protection context with information they can appreciate and understand. Legal icons have the potential to communicate legal norms including *obligation*, *prohibition*, and *permission* [6]. Pursuant to EU consumer protection and data protection law, CSPs must abide by a variety of norms at various levels. Using legal icons has the potential to provide CSPs with an alternative and accessible means of compliance. It also empowers users by providing a means of understanding key information regarding their rights under their contract with a CSP concerning how their data will be processed and used by a provider.

In the following article, we consider some of the core informational requirements present in both consumer and data protection law and evaluate visual tools designed to meet informational or other legal requirements by communicating legal concepts to consumers via icons. These icons, visualising legal concepts, have the benefit of providing CSPs with a means of communicating information to users on a non-textual basis. Although visualisation of legal concepts has many uses in the processes of contracting, this article primarily focuses on the use of icons to communicate contract and privacy policy terms to a consumer at the conclusion of an agreement [7].

The article has the following structure. First, we describe some of the main concepts and theory behind the visualisation of legal information. Second, we outline these requirements from the perspective of EU data protection law. Third, we describe informational requirements pursuant to consumer protection law. In each of the respective sections, the article explains the means used to present information visually and evaluates some limitations. Finally, we provide a conclusion evaluating positive and negative aspects of visualising legal information in consumer and data protection law. As a final note, this article focuses on cloud services contracted for by consumers on a non-nego-tiated basis, as opposed to business-to-business transactions.[1]

2   Visualising Legal Information

In the early days of computing, the interface between man[2] and machine was not graph-ical. Many will remember the days when written commands had to be input into the computer, which then, hopefully, complied by producing a new line of text on the screen. Users of today's graphical user interfaces may believe that this textual and mathematical representation of code has vanished. However, the human-readable programme code is merely hidden behind a graphical user interface that can be manipulated by clicking on icons for discs, documents, and apps. The details of that revolution are beyond the remit of this article, but it suffices to state that graphical representations have essentially simplified the utilisation of devices and made technology accessible to a far larger group of users.

Thus far, this revolution has primarily been limited to what Lessig [8] calls West Coast Code (i.e., computer code). In East Coast Code (i.e., legal code), there are far fewer examples of graphical user interfaces for legal information. Similarly, IT profes-sionals use a variety of graphical languages, such as the Unified Modelling Language [9], to visualise IT systems. These visual representations are of key importance, partic-ularly during systems design and analysis because fairly complex cases can be easily visualised, and computer code is not user-friendly or even comprehensible by most decision-makers. Again, there is a parallel between computer code and legal code: If it is useful to employ a graphical language to represent the functions of an IT system constituted by code, then perhaps we should not exclude the possibility that visualising legal code may also have some utility at some point in the future. Luckily, many exam-ples of graphical representation of legal information are readily accessible on the Internet, or even commonly known.

The representation of such information is far from uncommon. Perhaps the most obvious examples are traffic signs and lights that visually communicate binding legal rules to road traffic users. In addition, more comprehensive graphics, pictures, or combi-nations thereof (e.g., in comics) are created primarily to communicate normative infor-mation. For example, the *New York Street Vendor Guide* translates the most commonly violated rules regarding sales activities on the city's streets into easily accessible diagrams [10]. These diagrams illustrate the rules for vendors selling food, souvenirs, or other products in a much more accessible fashion than traditional textual code.

Moreover, there are examples of comics that illustrate the potential of visualising legal information [11–14]. Although comics often also include written text, they still contain a significant visual element that can play a key role in drawing attention to the

---

[1]   In the business-to-business context, many of the consumer protection rules described herein are inapplicable. Cloud services obtained by consumers are often of the Software as a Service (SaaS) variety, but are not limited as such and protections apply to Infrastructure as a Service (IaaS) and Platform as a Service (PaaS) deployment models.

[2]   This expression is meant to refer to humans, irrespective of their gender.

text or making it more easily accessible to the audience. A particularly nice example is a comic submitted as an *amicus curiae* brief to a US court [15], complete with references to relevant case law.

Visual representations of norms are by no means a new phenomenon. Historically, legal iconography has had an important function in communicating legal rules to illiterate people. Moreover, crime prevention has sometimes taken the task of general prevention to an exhibitionistic extreme.[3] Thus, there is no doubt that graphical representations can have a significant impact on the act of communicating legal information. Its functions include, but are not limited to, drawing attention to the legal information and communicating key aspects fairly quickly. However, visual representations have limitations because they often cannot achieve the same level of abstraction as a textual representation [14, 16]. Moreover, visual representations may be overly ambiguous when they are not based on a graphical language that is sufficiently well known and clear.

Legal icons, such as road traffic signs, are arguably the best examples of an existing graphical language that is universally understood. Road traffic signs are essentially icons that condense a rule into a singular graphic representation that can be recognised by everyone who has learned this visual language. Traffic education focuses on learning this graphical language, and an international convention ensures that the language is universally understood, based on a common legal iconography. These signs form part of internationally understood visual semiotics, which has spread beyond the context of road traffic to include other domains (e.g., no smoking signs) [17].[4] As will be shown in the next sections, similar icons could potentially be used in many other fields including data privacy and consumer protection domains.

3 Information Provision Under the EU Data Privacy Framework

3.1 Fairness of Processing

One of the core principles of the European Data Privacy framework is that personal data must be processed fairly. The fairness principle requires that processing operations are able to meet the reasonable expectations of the individuals. This principle ensures that the processing of personal data does not exceed the expectation of individuals and that its further processing is not objectionable in light of these expectations. Under the Data Protection Directive [18], the concrete application of the fairness principle is anchored in two main rules. The first requires the data controller (i.e., any entity processing personal data) to file a notification with the relevant national Data Protection Authority (DPA) before carrying out any wholly or partly automatic processing operation or set of such operations intended to serve a single purpose or several related purposes. This

---

[3] For example, the severed hand of a presumed criminal is still exhibited today in the town hall of the German city of Münster, constituting one of the more extreme symbols of legal risk.

[4] See Annex 1 of Vienna Convention on Road Signs and Signals.

allows data subjects consulting the national publicly accessible register to find out how their personal data are processed.

The other aspect of fairness requires the data controller to provide each individual with a minimum amount of information about the processing and the identity of the data controller (i.e., the transparency requirement). This right is part of the broader right to information self-determination and the right to information, which is a fundamental right under the EU legal framework. Furthermore, economists have always analysed the importance of information in making rational decisions in a market [19]. Information asymmetry between market players is considered to lead to inefficient transactions because the party without the relevant information cannot make a rational choice and might be forced to engage in other transactions to replace or remedy the defects of the initial inefficient transaction. In economics, the common remedy for such a problem is to force the party with information to disclose information relevant to the other party. The transparency rule embodied in the fairness principle reinforces the need to remedy the information asymmetry between the company who has the information about the processing and the individual that needs access to the information to enter into an efficient transaction. In this sense, this principle has a strong foothold in economic ideals and seeks to prevent potential market failures due to information asymmetry [20].

The focus of this section is the transparency requirement, which is at the centre of the fairness principle.[5] Article 12 of the Directive [18] aims to ensure the transparency of the processing of personal data by providing the data subject information regarding its purpose, recipients, retention period, and so forth. Compliance with this requirement would dictate having a policy for how the company treats the personal data. The transposition of this rule in some member states—for example, in the UK, —explicitly requires having a privacy policy [21]. Outside the EU, the US Federal Trade Commission (FTC) encourages companies to deploy privacy policies.

Often, data controllers try to comply with this requirement, as mandated or voluntarily, by adopting lengthy and bulky privacy policies that are hardly read or understood by data subjects. According to the 2015 Eurobarometer survey on data protection, only one in five respondents (18 %) fully read privacy statements [22]. These privacy policies are so lengthy that it would take an average person, according to one study, about 250 working hours (30 full working days) every year to read the privacy policies of the websites they visit [23]. Another study, from 2008, uses monetary value and estimates that the opportunity cost of reading the privacy policies could reach up to $780 billion dollars annually [24]. However, it is not only that such policies are often long and time-consuming. Even when one decides to read the policies, they are obscure and full of legalese. Leaving data subjects in the dark—i.e. confusology, as it is currently referred to in the literature—in terms of how their data is being used is becoming a prevalent business practise, by which firms 'purposefully introduce uncertainty and confusion into consumer transactions' [25].

---

[5] Recital 38 of the preamble of Directive 95/46/EC provides that 'if the processing of data is to be *fair*, the data subject must be in a position to learn of the existence of a processing operation and, where data are collected from him, must be given accurate and full information, bearing in mind the circumstances of the collection.'

Such practises present a significant challenge for the exercise of individual rights to privacy. In the absence of adequate and sufficiently understandable information regarding how data is processed, effective control of data by users becomes challenging and, as a result, data subjects are unable to make informed decisions about their data. This also makes it challenging for both regulators and individuals to hold the entities accountable. This, in turn, affects consumer trust in using digital services, signifying the need for systematic and innovative ways of communicating information to data subjects. Some of the changes under the EU data protection reform are aimed at mitigating these problems associated with the cognitive limitations of data subjects. A notable development to this end is related to the initiatives to standardise certain aspects of such information provision by introducing standards for communicating with data subjects in very concise and understandable manner, which is discussed in the next section.

3.2   Towards Iconised Privacy Policies

Unlike the Directive, the Regulation makes an explicit reference to the principle of transparency. Article 5(a) of the Regulation [26] provides that 'personal data shall be processed lawfully, fairly, and in a transparent manner in relation to the data subject'. Article 14 of the Regulation still contains an obligation to provide extensive information regarding the processing. It is even considered to increase the amount of information that must be provided to data subjects [27]. As noted above, the challenge with such information provision rules has always been to provide information to the extent and in a form useful to data subjects. In his work on information regulation, Cass Sunstein underlines that '[w]ith respect to information, less may be more' [28]. Furthermore, as the famous adage goes, 'a picture is worth a thousand words'. The Regulation attempts to standardise certain aspects of this information provision by introducing the possibility of using standard icons. Article 12(7) of the Regulation indicates that the information to be provided by the controller under Articles 13 and 14 'may be provided in combination with standardised icons'. Initially, the use of such visual icons was suggested by the European Parliament [29] and was an integral part of the Regulation. However, in the final draft of the Regulation [26], which is published in the official journal, the implementation of such standardised icons are left to the Commission's delegated act. Under Article 290(1) of the Treaty on the Functioning of the European Union (TFEU) [30], delegated acts are non-legislative acts that supplement or amend non-essential elements of a legislative act. Once enacted, and absent objection from the Parliament or the Council, delegated acts will have a binding effect [30]. Given that the market has been slow in taking the initiatives upon itself, the requirement under the Regulation would give a much needed push in the undertaking of such simplified and easy-to-grasp ways of communicating information. Similar initiatives by the EU to introduce mandatory standardised information provision through labelling in the energy sector have yielded encouraging and positive results in enhancing consumer understanding [21].

According to Article 12(8) of the Regulation [26], the Commission's delegated acts should elaborate what information needs to be presented by the icons and the procedures for providing standardised icons. It is not clear whether the Commission would adopt exactly the same icons and procedures as suggested by the Parliament, although it is reasonable to assume that the Commission would make use of the already existing work from the Parliament. However, it is important to note that the use of icons is not meant to replace the more detailed provision of textual information. Instead, it aims to complement the existing ways of providing information, as stipulated under Article 14 of the Regulation. As noted above, such icons are initially suggested by the Parliament, and it is this draft that contains more detailed information about the icons. Even though the icons as suggested by the Parliament did not make it to the final draft, it at least started the conversation and will likely feature in the Commissions' delegated works. Thus, in the following paragraphs, the visual icons and procedures suggested by the Parliament are highlighted, followed by brief discussion of their implications in the cloud computing context.

In this regard, apart from the information to be provided under Article 14, Article 13a of the suggestion from the Parliament [29] requires the controller to use of standardised icons in providing certain essential information regarding the underlying data processing operations. The information relevant for the standardised notice includes the following:

- whether personal data are collected beyond the minimum necessary for each specific purpose of the processing;
- whether personal data are retained beyond the minimum necessary for each specific purpose of the processing;
- whether personal data are processed for purposes other than the purposes for which they were collected;
- whether personal data are disseminated to commercial third parties;
- whether personal data are sold or rented out; and
- whether personal data are retained in unencrypted form.

According to Article 13a(2), such information should be presented 'in an aligned tabular format, using text and symbols' as shown in Fig. 1.

Both Article 13a and the Annex of the Parliament draft provide detail requirements in the implementation of such formats. Among other things, first, the order and format of the above notice cannot be changed by the controller [29]. As shown in Fig. 1, the first column should contain the icons, the second column should describe the icon and the third column should indicate whether that requirement is met. In providing the information in the second column, it must also be noted that some of the words have to be in bold (see more Annex [29]). Similarly, the third column must use the following two icons to indicate whether or not the requirement is fulfilled (Fig. 2):

ICON	ESSENTIAL INFORMATION	FULFILLED
	No personal data are **collected** beyond the minimum necessary for each specific purpose of the processing	
	No personal data are **retained** beyond the minimum necessary for each specific purpose of the processing	
	No personal data are **processed** for purposes other than the purposes for which they were collected	
	No personal data are **disseminated** to commercial third parties	
	No personal data are **sold or rented out**	
	No personal data are retained in **unencrypted** form	

COMPLIANCE WITH ROWS 1-3 IS REQUIRED BY EU LAW

Fig. 1. Standardised format for information provision (adopted from [29])

Fig. 2. Standard icons for marking compliance (adopted from [29])

This means that if, for example, the controller does not collect information "beyond the minimum necessary for each specific purpose of the processing', then the specific third column of the first row should look like as follows (Fig. 3):

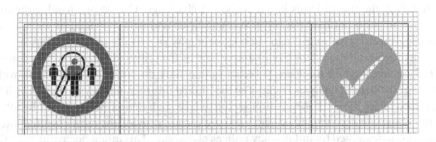

Fig. 3. Illustration of the standardised formats (adopted from [29])

Second, when such information is provided electronically, it should be clear, legible, and machine-readable.[6] At least the text in column two should also be provided in a language easily understandable by the consumers in the Member State. Third, the controller should not provide additional information or explanation within such formats. If desired, this should be done with the other information as provided under Article 14 [26]. Furthermore, the Annex provides more detailed requirements in terms of the size and width of the icons. Although such detailed guidance is important in ensuring that such icons are used uniformly by different actors processing personal data, it is not clear what would occur if one were to fail to adhere to technical aspects such as the size, width, and the order of the icons. Some work also needs to be done in terms of preventing the use of such icons for purposes other than providing privacy-related information in a way that might create confusion to consumers. In the offline world, a common way of dealing with this problem is to register the icons as trademarks or trade names so that their usage would be limited to privacy contexts.

Standardising the information provision rules is a move in the right direction, at least from the data subject's point of view. Research shows that the use of such icons facilitates an easy-to-grasp way of communicating complex information, thereby enhancing users' understanding of privacy policies [21]. Related to this, the use of such icons improves the potential for communicating information uniformly across all studied groups, regardless of their educational and cultural backgrounds [21]. Respondents in a focus-group-based study found icon-based policies to be 'clear and helpful' [21]. A more informed customer could only benefit the wider uptake of cloud services. The more confident consumers are in terms of what happens to their data, the more they can trust the cloud. The usefulness of such icons is not limited to communicating information from entities to customers. Cloud providers could improve their internal compliance

---

[6] The machine-readable requirement also comes from Recital 60 of the final draft of the Regulation. See also Article 13a(3) of the Parliament draft.

framework using industry-developed icons. For example, the Prime Life project and the PrivIcons have developed icons for email privacy, which could be used to ensure the confidentiality of email communications that contain customers' personal data or any confidential information [21]. There are also other icons that could be used to communicate ownership of information over cloud-stored data [31].

However, the use of icons is not without a challenge. On the one hand, assuming that most users do not read the full privacy policies and would depend on such icons, the question is to what extent can such iconography of privacy polices empower user decisions regarding their data. For example, one of the critics of earlier initiatives from the EU on labelling in the energy sector is that they do not allow consumers well-informed cost-benefit analysis because of the limited information they provide [21]. The suggestion under the draft Regulation seems to take such limitations on board by introducing a layered approach, where the icons are used to provide the most essential information, complemented by full privacy polices elsewhere. This might not avoid potential reliance on the icons, but it provides consumers with more detailed information if they are interested in digging deeper into how their personal data are treated.

More importantly, the fact that a controller has to indicate that it has met the requirement does not enhance actual compliance by controllers. Compliance with the standard notice might even lead to a tick-the-box approach and create a false sense of protection for the data subjects. This means that unless such schemes are complemented with an auditing and certification mechanism, they would have limited significance in enhancing the rights of individuals. In fact the Regulation has introduced certification mechanisms, and there are also on-going standardisation initiatives, which can be used to strengthen the use of icons. Similarly, how is the concern of visually impaired people addressed with regard to this standard notice? Does this obligation require the utmost effort from controllers to provide the necessary information in a manner that people can comprehend?

This obligation will affect many cloud providers that have individual users as customers. Those providers will be required to provide information regarding the processing of information in this standard notice. This includes the implementation of a machine-readable format of the standard notice requirements when the information is provided electronically. This means, for example, websites providing privacy policies should implement certain standards, including standards to address readability on different devices such as mobiles. This implies the need to maintain two sets of privacy policies: one presented through icons and one providing more detailed information that is consistent with the information presented through the icons. Furthermore, companies often change their policies from time to time, underlying the need for updating the information provided through icons. There is no doubt that this would be a time-consuming and costly for cloud providers.

Overall, the initiative to encourage the use of icons to provide information is commendable. However, it can only improve the conveyance of providers' exiting policies, not their actual commitment to ensure the privacy of customers. If a real change in terms of users' rights is to come, it would require the willingness to consider adherence to privacy principle rules as a competitive advantage, rather than as a matter of compliance. In the next section, we consider a slightly varied approach to visualising law through icons at the European level in dealings with consumer contracting.

4   EU Consumer Protection Law: Application and Informational Requirements

In this section, we evaluate core EU rules for consumer protection and evaluate a recent attempt to make these rules more accessible to consumers by representing core aspects with icons. At the outset, we note that this system has promise. However, like the graphical user interfaces that are visible in place of programme code, what lies behind is often complex. We evaluate how these complexities might impact the overall effectiveness of the message the icons attempt to convey.

In the EU, an inclusive and multifaceted system of rights in consumer transactions is designed to provide European consumers with perhaps the most expansive level of consumer protection available globally. On that basis, consumers ought to be able to extend this expectation—that their rights are protected—to purchases made in the digital marketplace on a national, and even on an international, basis. This wide-ranging coverage is achieved by offering consumers remedies at several stages or levels of the contracting process in addition to making certain unfair terms offered by sellers unenforceable. By creating a 'floor' or minimum standard that allows consumers to disaffirm contracts based on subjective dissatisfaction, or even 'buyer's remorse', the European consumer has substantial rights and remedies when they enter into contracts online. These rights are expressed in a series of directives and regulations.

Central EU directives currently in place to protect consumers include the Unfair Terms Directive (UTD), the Unfair Commercial Practises Directive, and the Consumer Rights Directive (CRD), among others [32–34]. In addition to consumer-specific legislation, the Rome I Regulation (law applicable in contractual matters) and the Brussels I Regulation (jurisdiction) also have consumer-specific provisions [35, 36]. Additionally, the Electronic Commerce Directive (ECD) provides a framework for harmonising or providing consistent rules for online transactions—ultimately contributing to consistency in e-commerce across European member states [37]. Application of these rules cover the entire duration of a consumer contract from the advertisement of a service, to the contract *offer* and *formation* of a contract, through procedural and substantive issues regarding the content of terms, and finally setting the rules governing the *how* and *where* disputes will be adjudicated if the need arises.

As consumer protection regulations are contained in many different instruments, it can be difficult for consumers to understand and appreciate the rights they have. At the same time, less sophisticated CSPs and other providers—particularity those without legal counsel—may struggle to meet all of the requirements. As cloud computing is provided on a global scale, foreign providers are also required to conform to EU legal requirements, even if it is difficult for them to provide European consumers with the rights they have in their member states. For example, many of the contract terms offered by US-based CSPs are at odds with mandatory European consumer protection legislation described above, from price and informational terms to choice of law and forced arbitration requirements [38]. The following subsections provide some of the requirements pursuant to EU consumer protection law followed by evaluation of a graphical means, expressed through icons, designed to communicate requirements. Although the icons do not represent all aspects of consumer protection law, they might providers with a basis for complying with principal informational requirements, among others.

4.1   Name Address and Contact Details

Both the CRD and the ECD require that the seller provide information regarding the name of the trader or service provider [34, 37]. The CRD and the ECD also require that the seller provide contact information. Although the requirements are similar, the CRD requires that the seller provide 'the geographical address at which he is established and his telephone number'. The ECD requires similar information, but does not require that a telephone number be provided (i.e., email address is sufficient).

4.2   Costs, Technical Requirements, and Product Information

If the cost of using a distance communication is above a basic rate, the CRD requires that that cost be communicated to the consumer [34]. In addition to cost, the ECD requires that the seller provide the technical steps needed to conclude the agreement, whether or not the contract will be filed by the service provider, technical means for correcting or rectifying errors in the order, languages available, and receipt of order by electronic means, without undue delay [37]. The CRD requires additional information regarding the *functionality* and *interoperability* of the product being offered [34]. Specifically, the seller must present 'any relevant interoperability of digital content with hardware and software that the trader *is aware of* or *can reasonably be expected to have been aware of*' [34]. This puts the burden on the CSP to be active and provide the required information.

4.3   Price

Both the ECD and the CRD have rules regarding price information to be provided to consumers. The ECD requires that the price term be 'indicated clearly and unambiguously' to the consumer [37]. The CRD requires clear information on the total price of the service, including taxes and any other charges that can be calculated [34]. According to the guidance document published on the CRD, the seller must notify the consumer of the total cost per billing period and the total monthly costs [34]. The point of departure is to prevent the many 'hidden' charges often associated with digital products and to reduce misrepresentation of prices.

4.4 Contract Information: Duration and the Right to Withdraw

The CRD requires that information regarding the duration of the contract, renewal requirements (i.e., automatic extensions of the agreement), and the consumer's obligations under the agreement are provided [34]. Cloud contracts are often provided on a monthly or yearly subscription basis, without a set duration. Therefore, under the CRD, the CSP must provide the consumer with information regarding conditions for terminating the contract [34]. The CRD also requires that consumers be provided with certain information regarding their right to withdraw from a service, including application of the right, procedures for exercising the right, the consumer's obligations for the costs of returning goods, and the obligation of the consumer to bear the traders' reasonable costs [34, 39].

4.5 Codes of Conduct

In the EU, there has been a push to create codes of conduct and other 'self-regulatory' initiatives to provide consumers with better information on products and to provide traders or sellers with guidance on how to comply with applicable laws. The starting point of these initiatives may be self-regulatory (set by the industries to be regulated) or co-regulatory (set by the industry with input from governments and regulators). The point of departure is that many instruments, such as codes of conduct and so-called 'trust seals, are voluntary, but members may have certain conditions imposed on them as a requirement of membership'. If members fail to meet those conditions, their actions may have consequences. For example, if a trader has earned the right to use a trust seal, and they act in a manner incompatible with the principles for using that seal, the industry organisation may require that the seal be removed or even 'black-listed' the seller from further use of the seal.

One problem with codes of conduct is they are often difficult for consumers to find and pair with the products they are purchasing [40]. To address this problem, the CRD requires the seller to inform the consumer of relevant codes of conduct where applicable and provide information on how copies of such codes can be obtained. The ECD provides a similar requirement, albeit a more limited one requiring only that '[a]ny relevant codes of conduct to which he [the trader] subscribes and information on how those codes can be consulted electronically' [37].

4.6 Visualisation of Contract Terms

A unique aspect of the CRD is an 'optional model' that sellers may adopt for displaying contract terms visually. Visualisation, or adding supplemental information to a contract in the form of icons or other information, potentially provides a clearer means for communicating legal requirements. Much like traffic signs represent laws or regulations —such as the direction of traffic or parking restrictions—icons or symbols in a contract might also be used to communicate legal concepts or principles in a much more user-friendly manner. For example, the following icons attempt to provide consumers with accessible points of reference to access information (Fig. 4):

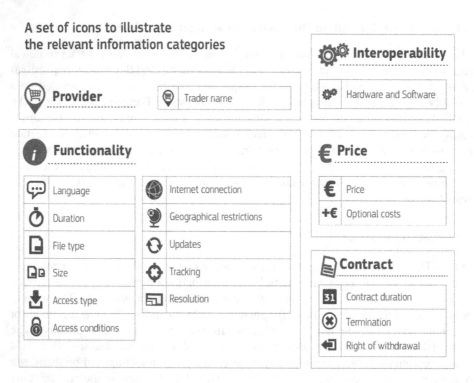

A set of icons to illustrate
the relevant information categories

Provider · · · · · · · · · · · · · · · ·   Trader name

Interoperability · · · · · · · · · · · · · · · ·   Hardware and Software

Functionality · · · · · · · · · ·

Language	Internet connection
Duration	Geographical restrictions
File type	Updates
Size	Tracking
Access type	Resolution
Access conditions	

€ Price · · · · · · · · · · · · · · · ·

| € | Price |
| +€ | Optional costs |

Contract · · · · · · · · · · · · · · · ·

31	Contract duration
⊗	Termination
←	Right of withdrawal

Fig. 4. CRD 'Optional Model' (adopted from [41])

There are some clear advantages to displaying legal information in this manner. For example, icons are much easier for consumers to read on a mobile screen than are dense contracts or privacy policies. If consumers understand the icons and the icons are consistently used, consumers will have a pretty good idea of where they need to look in order to obtain the information relevant for their purchase. To some extent, the icons also avoid the problem of presenting consumers with a great deal of complex information right before concluding a contract. Even if the terms are not fully understood, the icons potentially provide a much greater opportunity for at least a basic understanding of the contract terms, compared to providing a 30,000 word document alone for a mobile application.

Sellers adopting these tools may also be taking an important step towards providing less ambiguous contract terms, as required by several consumer protection instruments [32]. Although sellers are not required to use this model to comply with the information requirements of the CRD Article 6, the icons provide a readily available tool [41]. CSPs may adopt their own means of displaying legally operative information. However, the model provided in the CRD Annex provides a possible path for presenting information that will not require new development on the part of CSPs. Additional icons might aid in presenting general information, such as trader name, legal information including 'termination', 'contract duration', and technical information that impacts use of the service, in addition to payment and even certain privacy implications (i.e., 'tracking'). The following icons provide potential for expressing many contract terms:

Although the icons above, represented in Fig. 5 provide a potential step forward in providing consumers with complex information in a form that they are more able to understand, there are also clear limitations. Most users of cloud services are not lawyers. Without training or education, they will likely have difficulty understanding the rights represented by the icons. Like drivers are educated to recognise traffic signs, consumers will need to be educated and learn the visual language of the CRD model. Although somewhat intuitive, many of the rights represented by icons are far from obvious. For example, looking at Icon 17, it is not immediately apparent that the icon represents 'resolution'. Similarly, the arrow and box in Icon 18 represents the right to withdraw from a contract, but it could easily have many other meanings, including portability. Although others may be clearer, such as the lock used in Icon 7 or the price in Icon 15, the exact meaning and legal implications requires a deeper knowledge of the concepts represented. Building familiarity and understanding of the icons among consumers is therefore a crucial step to ensuring their effectiveness.

Set of Icons	01	Trader Name	08	Acces type	15	Price
	02	Language	09	Termination	16	Optional costs
	03	Duration	10	Updates	17	Resolution
	04	Internet connection	11	File Type	18	Right of withdrawal
	05	Geographical Restrictions	12	Size	19	Video on demand
	06	Tracking	13	Contract duration	20	Weather application
	07	Access conditions	14	Hardware and Software	21	Music application

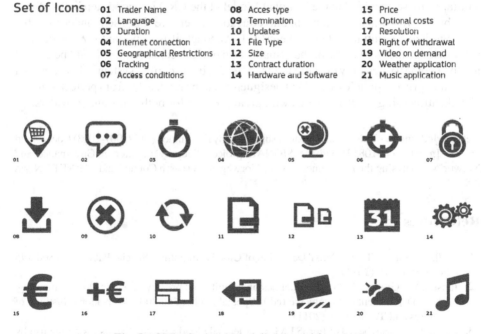

Fig. 5. CRD 'Optional Model' (adopted from [41])

5   Conclusion

Even if consumers do not read or understand complex contract terms and privacy policies, they still care about whether they are being treated fairly in the contracts they enter and that their privacy is protected. Exploring alternative methods for delivering vital information to consumers is a move in the right direction towards providing more complete notice and obtaining meaningful consent. Even if the current EU suggestions and icon models are not obligatory, they mark significant progress in the advancement of these ideas.

At the same time, visualisation of legal concepts is not without challenges—for example, oversimplification of or inadequate communication around nuanced and complex data protection or contract principles. Such concepts are difficult to convey with a picture. Where conveyed, such representations may oversimplify data collection practises or fail to convey the breadth of the fundamental rights that data subjects or consumers have in the EU. Similarly, there are concerns that although visual expressions may increase access for users with literacy impairment, such visual expressions could also potentially overlook the interests of the visually impaired. Making certain that users are not left behind is a difficult but important task.

Although some terms, like price or duration, are easy to express, more abstract principles, such as 'fair and lawful processing of data' will remain difficult. Perhaps the road forward on the data protection front involves providing additional options or a greater selection of symbols for providers to choose from. Conceivably, an expansion of icons similar to those provided in the Optional Model of the CRD is a good step in this direction, but it would again require public education to ensure that users understand the message being communicated to them [41]. Although it is important to consider and acknowledge the limitations of the visualisation of legal concepts, we should be careful not to become fixated only on these limits. After all, in the current system, where contract terms and privacy policies are all but designed not to be read, is far from perfect. Greater visualisation using icons is an area with great promise for both users and providers.

Acknowledgment. This work was partly supported by EU-funded (FP7/2007-2013) Coco Cloud project [grant no. 610853] and the SIGNAL project (Security in Internet Governance and Networks: Analysing the Law) funded by the Norwegian Research Council and UNINETT Norid AS.

References

1. Mell, P., Grance, T.: The NIST Definition of Cloud Computing (Special Publication 800-145 edn., Version 15 (2011)
2. Jansen, W., Grance, T.: NIST guidelines on security and privacy in public cloud computing. In: U.S. Department of Commerce (ed.) (Special Publication 800-144: National Institute of Standards and Technology (2011)
3. Reinecke, P., Seybert, H.: EuroSTAT Internet and cloud services - statistics on the use by individuals (2014)
4. Waelde, C., Edwards, L.: Law and the Internet, 3rd edn. Hart Publishing, Oxford (2009)
5. Matwyshyn, A.M.: Privacy the hacker way. Southern California Law Review, vol. 87(1) (2013)
6. Mahler, T.: Visualisation of legal norms. In: Jon Bing: En Hyllest/A Tribute. Gyldendal Norsk Forlag A/S, pp. 137–153 (2014). ISBN: 9788205468504
7. Barton, T.D., Berger-Walliser, G., Haapio, H.: Visualization: seeing contracts for what they are, and what they could become. J. Law Bus. Ethics **19**, 47–64 (2013)

8. Lessig, L.: Code version 2.0 (Basic Books) (2006)
9. Rumbaugh, J., Booch, G., Jacobson, I.: The Unified Modeling Language Reference Manual, 2nd edn. Addison-Wesley, Boston (2004)
10. Chang, C.: Street Vendor Guide: Accessible City Regulations (2009)
11. Hilgendorf, E.: Beiträge zur Rechtsvisualisierung (Logos) (2005)
12. Röhl, K.F., Ulbrich, S.: Recht anschaulich: Visualisierung der Juristenausbildung. Halem, Köln (2007)
13. Hoogwater, S.: Beeld"al voor juristen: Grafische modellen om juridische informatie toegankelijker te maken (Boom Juridische uitgevers) (2009)
14. Brunschwig, C.: Visualisierung von Rechtsnormen legal design (Schulthess) (2001)
15. Kohn, B.: Amicus Curiae, Brief to the United States District Court for the Southern District of New York
16. Brunschwig, C.: Tabuzone juristischer Reflexion, Zum Mangel an Bildern die geltendrechtliche Inhalte visualisieren. In: Schweighofer et al. (ed.), Zwischen Rechtstheorie und e-Government, Aktuelle Fragen der Rechtsinformatik (2003)
17. Wagner, A.: The rules of the road, a universal visual semiotics. Intl. J. Semiotics Law **19**, 311–324 (2006)
18. Directive 95/46/EC of the European Parliament and of the Council of 24.10.1995 on the protection of individuals with regard to the processing of personal data and on the free movement of such data. OJ L 281/31
19. Posner, R.A.: Economic Analysis of Law, Aspen Casebook Series, 8th edn, Aspen Publishers, New York (2011)
20. Lynskey, O.: Deconstructing data protection: the "added-value" of a right to data protection in the EU legal order. Intl. Comp. Law Q. **63**(03), 569–597 (2014)
21. Edwards, L., Abel, W.: The use of privacy icons and standard contract terms for generating consumer trust and confidence in digital services. CREATe working paper series. 10.5281/zenodo.12506
22. Special Eurobarometer 431, Data Protection (European Commission, 2015) Catalogue Number DS-02-15-415-EN-N
23. World Economic Forum, Unlocking the Value of Personal Data: From Collection to Usage (2013)
24. McDonald, A., Cranor, L.: The cost of reading privacy policies. In: Proceedings of the Technology Policy Research Conference, 26–28 September 2008
25. Calo, R.: Digital market manipulation. Geo. Wash. L. Rev. **82**, 995 (2013)
26. Regulation (EU) 2016/679 of the European Parliament and of the Council of 27 April on the protection of individuals with regard to the processing of personal data and on the free movement of such data, and repealing Directive Directive 95/46/EC (General Data Protection Regulation, hereinafter GDPR) OJ L 119/1
27. De Hert, P., Papakonstantinou, V.: The proposed data protection Regulation replacing Directive 95/46/EC: a sound system for the protection of individuals. Comput. Law Secur. Rev. **28**, 130–142 (2012)
28. Sunstein, C.R.: Information regulation and information standing: akins and beyond. University of Pennsylvania L. Rev. **147**, 613 (1999)
29. Committee on Civil Liberties, Justice & Home Affairs, Report on the proposal for a regulation of the European Parliament and of the Council on the protection of individuals with regard to the processing of personal data and on the free movement of such data, 21 November 2013. Available at (LINK) (Hereinafter Parliament Draft)
30. Consolidated Version of the Treaty on the Functioning of the European Union, Article 289(1) (2012) O.J (C 326)

31. Helton, A.: Privacy Commons Icon Set (2009). http://aaronhelton.wordpress.com/2009/02/20/privacy-commons-icon-set/
32. Council Directive 93/13/EEC of 5 April 1993 on unfair terms in consumer contracts, OJ L 095, 21/04/1993, pp. 0029–0034 (Unfair Terms Directive)
33. Directive 2005/29/EC of the European Parliament and of the Council of 11 May 2005 concerning unfair business-to-consumer commercial practices in the internal market and amending Council Directive 84/450/EEC, Directives 97/7/EC, 98//27/EC and 2002/65/EC of the European Parliament and of the Council and Regulation (EC) No 2006/2004 of the European Parliament and of the Council, OJ L 149, 11/06/2005, pp. 0022–0039 (Unfair Commercial Practices Directive)
34. Directive 2011/83/EU of the European Parliament and of the Council of 25 October 2011 on consumer rights amending Council Directive 93/13/EEC and Directive 1999/44/EC of the European Parliament and of the Council and repealing Council Directive 85/577/EEC and Directive 97/7/EC of the European Parliament and of the Council Text with EEA relevance (Consumer Rights Directive)
35. Regulation (EC) No. 593/2008 of 17 June 2008 on the law applicable to contractual obligations (Rome I)
36. Council Regulation (EC) No. 44/2001 of 22 December 2000 on jurisdiction and the recognition and enforcement of judgments in civil and commercial matters (Brussels I)
37. Consumer Rights Directive Art. 3. See also Rec. 22. The ECD introduces concepts such as the 'country of origin rule' to harmonize the rules (licensing etc.) that online actors must comply with. Essentially, this requires that CSPs only have to follow the regulations of the country where they are established, not the rules of all member states
38. Rustad, M.L., Onufrio, M.V.: Reconceptualizing consumer terms of use for a globalized knowledge economy. Univ. Pennsylvania J. Bus. Law **14**, 1085 (2012)
39. Millard, C.J.: Cloud Computing Law. Oxford University Press, Oxford (2013)
40. Loos, M.B.M.: Analysis of the applicable legal frameworks and suggestions for the contours of a model system of consumer protection in relation to digital content contracts. University of Amsterdam (2011)
41. European Commission, Optional Model. http://ec.europa.eu/justice/consumermarketing/files/model_digital_products_info_complete_en.pdf. Accessed 3 Nov 2015

Brief Overview of the Legal Instruments and Restrictions for Sharing Data While Complying with the EU Data Protection Law

Francesca Mauro[✉] and Debora Stella

Studio Legale Bird & Bird, Milan, Italy
{francesca.mauro,debora.stella}@twobirds.com

Abstract. Data are the new oil of our society, but as opposed to the latter, business are not allowed to work them and re-use freely. To the extent that data fall under the category of "personal data", businesses must comply with the data protection legal framework. In order to do this, it is primarily necessary to design internal and automatic procedures to understand if the sharing of data, as further processing operation, is compatible with the original purpose, and if appropriate safeguards – such as anonymisation – can be implemented without compromising achievement of the aim pursued through the sharing. When the aim of the sharing requires businesses to disclose personal data, businesses must detect a legal ground to rely upon and comply with several data protection rules. The aim of this paper is to briefly analyze solutions adopted by stakeholders under the EU data protection legal framework.

Keywords: Data sharing · EU data protection law · Purpose limitation · Data minimization · Anonymised data · Data subjects' rights · Privacy by design

1 Introduction

Nowadays the market offers a wide range of activities which may result in online sharing of data of whatsoever nature and for any reason: from the open data government initiative to the several projects for sharing research data among the scientific community, from the social networks phenomenon to the cloud services.

Sharing data has not only an economic value, but is fundamental for the progress of mankind and of a data driven economy, a priority also recognized by the Digital Single Market Strategy of the European Commission [1, 2].[1] On the other side, however, in

---

[1] For a complete overview on the value of the sharing economy in the European Union see European Parliamentary Research Service "The Cost of Non-Europe in the Sharing Economy" (2016), according to which the new data protection legislation (i.e. Regulation (EU) 2016/679 of the European Parliament and of the Council of 27 April 2016 on the protection of natural persons with regard to the processing of personal data and on the free movement of such data - General Data Protection Regulation) will help the development of the sharing economy by enabling citizens to exercise effectively their rights to personal data protection and modernizing and unifying rules for businesses to make the most of the Digital Single Market.

© Springer International Publishing AG 2016
S. Casteleyn et al. (Eds.): ICWE 2016 Workshops, LNCS 9881, pp. 57–68, 2016.
DOI: 10.1007/978-3-319-46963-8_5

certain circumstances, sharing of data can jeopardize fundamental rights of individuals as the information may be used to discriminate individuals by refusing to provide certain persons with services because of their health, religious or economic status [3]. It can also be used in a way, or for purposes, which can affect individuals' dignity and last, but not least, violate the individuals' right to respect their private life and personal data. [2] Control of individuals through their personal data remains one of the main topics and challenges to be addressed and solved when free flow of data is a priority of the new economy. Toward this goal the European Commission recently announced its new strategy in the Digital Single Market, including the adoption of a future-proof legislation that will support the free flow of data [4].

For the purpose of this paper it is worth specifying that there is no definition of "sharing of data" under the EU Data Protection Law, including the Regulation (EU) 2016/679 of the European Parliament and of the Council of 27 April 2016 on the protection of natural persons with regard to the processing of personal data and on the free movement of such data. Some EU data protection authorities, which have provided guidance on data sharing agreements, define "data sharing" as the "disclosure of the data by transmission, dissemination or otherwise making it available" in many different contexts, i.e. within the public or private sectors, or among the public and/or private organizations [5].

In practice, "data sharing" is commonly used by operators when referring to different activities which can be classified under two specific categories of data processing operations explicitly mentioned by the EU Data Protection Law: "disclosure" and "dissemination".

In particular, where data are communicated, released or in any other way made available to a limited number of individuals (i.e. identifiable recipients), this is regarded as "disclosure". Conversely, "dissemination" usually occurs when personal data are spread among an indefinite number of unknown persons.[3] Keeping in mind the difference between disclosure and dissemination can be worthwhile when speaking about online sharing: sharing information through the web may be subject to restrictions that apply to dissemination only.

By way of example, the share of information carried out by governmental bodies in the context of the open data initiative or a data base freely accessible online from anywhere and by anybody in the world can be regarded as "dissemination". On the other

---

[2] The right to data protection is a fundamental right in EU Law. It is established under Article 8 of the Charter of Fundamental Rights of the European Union which became legally binding as EU primary law with the coming into force of the Lisbon Treaty on December 1, 2009. Even if fundamental, the right to the protection of personal data is not an absolute right, but must be considered in relation to its function within society. Therefore, a balancing exercise with other rights (e.g. freedom of expression, access to document, freedom of the arts and sciences) is necessary when applying and interpreting the right to data protection.

[3] The distinction between "disclosure" and "dissemination" is important for some data processing operations. By way of example, under Article 26 of the Italian Data Protection Code (Legislative Decree 30 June 2003 n. 196) disseminating health data is prohibited while their disclosure, under certain conditions, is permitted.

hand, where information is exchanged between two organizations through, e.g. a data room available in cloud, this is more correctly referenced as "disclosure" of data.

Irrespective of whether data sharing is disclosure or dissemination, the fact remains that sharing personal data constitutes a "personal data processing operation," a condition that triggers the application of the EU Data Protection Law.[4]

The goal of this paper is primarily to introduce the reader to the data protection legal framework which applies when a sharing of personal data occurs. Data controllers who are about to share information that can fall under the definition of "personal data" should first conduct a so called "data protection impact assessment" in order to identify roles of the parties involved in the sharing, and assess the purposes –we and possible related risks – arising from this sharing of personal data. Furthermore, since sharing personal data may also imply a re-use of the personal data for purposes other than the original one(s), data controllers need to find a legal ground to rely upon. The analysis continues by focusing on the main solutions usually adopted by businesses that decide to share their data. On the one hand, there are businesses which implement de-identification techniques for the purpose of avoiding the application of the EU Data Protection Law; this paper tries to highlight the related risks. On the other hand, there are businesses that cannot avoid sharing data in the form of "personal data" for the purposes for which the data have been collected and, therefore, they have no choice: they must comply with the data protection legal framework, in which case the technology may help to facilitate these businesses to implement systems according to the privacy by design principle as described below.

The structure of this paper is as follows: Section 2 illustrates the limitations to data sharing deriving from the "purpose limitation principle" and provides criteria for assessing the compatibility analysis; Sect. 3 is intended to describe main current trends implemented by business to share the data; Sect. 4 describes the conclusions.

---

[4] Regulation (EU) 2016/679 provides two scopes of application: material and territorial. According to the material scope (Article 2), the Regulation applies to the processing of personal data wholly or partly by automated means and to the processing, other than by automated means, of personal data which form part of a filing system or are intended to form part of a filing system. Pursuant to the territorial scope, the Regulation applies to (i) the processing of personal data in the context of the activities of an establishment of a controller or a processor in the EU, regardless of whether the processing takes place in the EU or not; (ii) to the processing of personal data of data subjects who are in the EU by a controller or processor not established in the EU, where the processing activities are related to: (a) the offering of goods or services, irrespective of whether a payment of the data subject is required, to such data subjects in the EU; or (b) the monitoring of their behaviour as far as their behaviour takes place within the EU; (iii) to the processing of personal data by a controller not established in the EU, but in a place where Member State law applies by virtue of public international law.

2   Data Reusability and the "Purpose Limitation" Principle

The aims which can lead data controllers[5] to share personal data are countless: a large amount of private organizations regularly disclose personal data for executing contracts, fulfilling obligations provided under applicable laws or for the purpose of scientific researches. Other businesses collect personal data with the specific purpose to sell them to other private organizations for these organizations to use the data for their own business purposes. In this respect, however, it is worth to underline that often the disclosure or the dissemination is a subsequent phase of a processing of personal data in the data life-cycle. Data are indeed mainly collected for purposes other than the disclosure to third parties (unless in case of a business that collects data for the sole purpose of reselling them). By way of example, healthcare providers or telecommunication companies daily collect a large amount of personal data, respectively, of their patients and customers for the main purpose of providing them with their services. Nevertheless such providers may subsequently want, or be obliged, to disclose the above data for a wide range of reasons.

Given the above, sharing of data for purposes other than those originally collected may be considered as a re-use of data or, more precisely, in accordance with the data protection terminology, as a "further processing".

Now, according to one of the pillars[6] of the EU Data Protection Law (the *purpose limitation principle*) personal data must be processed only for specified, explicit and legitimate purposes (*purpose specification*) and must not be further processed in a manner that is incompatible with those purposes (*compatible use*) [6].[7]

---

[5] Under the Regulation (EU) 2016/679 a "data controller" is the natural or legal person, public authority, agency or other body which, alone or jointly with others, determines the purposes and means of the processing of personal data.

[6] The EU Data Protection Law permits to process personal data only under specific and limited circumstances (i.e. legal grounds) and requires data controller to comply with the following principles: lawfulness, fairness and transparency (i.e. data must be processed lawfully, fairly and in a transparent manner); purpose limitation; data minimization (i.e. data must be adequate, relevant and limited to what is necessary in relation to the purposes for which they are processed); accuracy (i.e. data must be accurate and, where necessary kept up to date); storage limitation (i.e. data must be kept in a form which permits identification of data subjects for no longer than is necessary for the purposes for which the personal data are processed); integrity and confidentiality (i.e. data must be processed in a manner that ensures appropriate security of the personal data).

[7] "Specification of purpose is an essential first step in applying data protection laws and designing data protection safeguards for any processing operation. Indeed, specification of the purpose is a pre-requisite for applying other data quality requirements, including the adequacy, relevance, proportionality and accuracy of the data collected and the requirements regarding the period of data retention. The principle of purpose limitation is designed to establish the boundaries within which personal data collected for a given purpose may be processed and may be put to further use." (Article 29 Working Party, WP203, p. 4).

2.1 The Compatibility Test

Compatibility needs to be assessed on a case-by-case basis through a substantive compatibility assessment of all relevant circumstances taking into account specific key factors based on the guidance given under Article 6, paragraph 3.a of the Regulation (EU) 2016/679 that can be summarized as follows:

- any link between the purposes for which the personal data have been collected and the purposes of the further processing: this may also cover situation where there is only a partial or even non-existent link with the original purpose;
- the context in which personal data have been collected and the reasonable expectations of data subjects as to the further use of their personal data. In this case the transparency about the use of the data originally reached by the data controller when it collected the data is of paramount importance: the more expected is the further use, the more likely it is that it would be considered compatible. In order to assess the reasonable expectation of individuals as to the use of their data, attention should be given also to the environment and context in which data are collected (i.e. the nature of the relationship between data controller and data subjects could raise reasonable expectation of strict confidentiality or secrecy, as it is usually in the healthcare-patient, or bank-account holder, relationships);
- the nature of personal data and the impact of the further processing on data subjects: particular attention must be paid when the re-use involves special categories of personal data such as the sensitive ones[8] as well as in case of biometric, genetic or location data and other kinds of information requiring special safeguard (e.g. personal data of children);[9]
- the safeguards adopted by the controller to ensure fair processing and to prevent any undue impact on data subjects: this factor may be, probably, the most important one because under certain circumstances it can help businesses to compensate for a change of purpose when all the other factors are deficient. In particular, in addition to appropriate"technical and organizational measures to ensure functional separation" (e.g. partial or full anonymisation, pseudonymization and aggregation of data), the data controller must have implemented "additional steps for the benefit of the data subjects such as increased transparency, with the possibility to object or provide specific consent" [6].

However, the newly approved Regulation (EU) 2016/679 recognizes the possibility for the data controller to avoid the assessment on the above factors for ascertaining the compatibility of the further processing if the controller can rely on the specific consent

---

[8] Regulation (EU) 2016/679 provides additional safeguards for the processing of "special categories" of personal data which are, by their nature, particularly sensitive in relation to fundamental rights and freedoms (i.e. data revealing racial or ethnic origin, political opinions, religious or philosophical beliefs, or trade-union membership, and the processing of genetic data, biometric data for the purpose of uniquely identifying a natural person, data concerning health or data concerning a natural person's sex life or sexual orientation).

[9] Regulation (EU) 2016/679 provides a special care for children by introducing limits and additional requirements when the processing concerns data of children under 16 years.

of data subject for using their personal data for further processing or in case the further processing is mandated by a legislative or statutory law to which the data controller must comply.

Processing of personal data in a manner incompatible with original purposes of the collection infringes the EU Data Protection Law and thus, is prohibited.

A key concept introduced by the EU Data Protection (Recital 40, Regulation (EU) 2016/679) is a "presumption of compatibility" with the initial purposes of the collection in the event that the further processing is carried out for achieving "purposes in the public interest, scientific or historic research purposes or statistical purposes", provided that appropriate safeguards are adopted for the rights and freedoms of data subjects in accordance with the principle of minimization.[10]

It should be finally clarified that, whenever personal data are shared with a third party service provider acting, and appointed by the data controller, as data processor,[11] this cannot be considered as a further processing for a new purpose: indeed, this disclosure of personal data occurs in order to fulfill the purposes of the collection (e.g. disclosing personal data of employees to a payroll provider appointed as data processor by the employer does not need a compatibility assessment since it is carried out for the purpose of executing the employment contract which is the original purpose of the collection of employees' personal data; similarly, the use by the data controller of cloud based services for processing operations related to the contract with customers does not technically trigger any "disclosure" or even a "further processing" provided that the service provider acts under the instructions of the data controller as data processor).

2.2   The Roles of the Parties

Given the above, it is therefore preliminarily necessary to identify roles and purposes of the disclosure in order to assess whether the sharing of data is with a third party that will act as data controller.

In the event that a business intends to disclose data to a third party who uses them autonomously (as a data controller), it would be required to assess the nature of the data and the purposes of their collection in order to evaluate whether the new purposes for which personal data are disclosed may be considered compatible with the original purposes. To this extent, a data protection impact assessment (and/or the implementation of privacy-preserving technologies, if applicable) could be worth to adequately set technical procedures. This is not required if the data controller has

---

[10] Under Article 89 of the Regulation (EU) 2016/679 pseudonymization is suggested as an appropriate safeguard when the processing of personal data occurs for archiving purposes in the public interest, scientific or historical research purposes or statistical purposes, without prejudice to other technique which may be more effective.

[11] Under the Regulation (EU) 2016/679 a "processor" is the natural person or the entity that processes personal data on behalf of a controller. By way of example, where IT services are provided by a third party service provider, the organization using the IT services offered by the third party appoints the IT provider as its data processor.

obtained the specific consent of data subjects (or the disclosure of personal data to the third party is required by laws).

One of the most important queries that a system should be able to formulate and to which a system should be able to respond is to what extent the sharing of "personal data" - i.e. any information relating to an identified individual or to an individual who can be identified also by reference to an identifier such as a name, an identification number, location data, an online identifier or to one or more factors specific to the physical, physiological, genetic, mental, economic, cultural or social identity of that natural person (Article 4, Regulation (EU) 2016/679) – is necessary for achieving the aim pursued through the sharing: in many cases businesses do not really need to share information that can be linked to an individual, i.e. personal data, because the scope for which the information is shared can be achieved also using anonymous data. Appropriate safeguards such as the implementation of anonymisation techniques may help businesses to comply with data protection laws and, at the same time, enabling them to make the necessary data available for the sharing [7].

3 Current Trends for Sharing Data

When the sharing of data is with a third party that receives the data to process them autonomously for its own purposes, depending on the answer given to the above queries, businesses can follow two different ways:

- they may choose to release data in a sufficiently aggregated (or even effectively anonymized) form in order to strengthen the protection of the individuals to whom data relate (this would partially simplify the related data protection obligations); or
- they may need to disclose the information as "personal data" and thus they should design and adopt technical solutions in order to comply with further data protection rules such as the provisions of smart mechanism to give, and also withdraw, the consent (if required) or to opt-out (if applicable), and implement tools for improving the data subjects' control over their data and simplifying the fulfillment of data subjects' request relating to the exercise of their rights.

3.1 Sharing De-Identified Data

Anonymization. The EU Data Protection Law does not apply to "anonymous data"[12] or to "personal data rendered anonymous in such a manner that a data subject is not, or is no longer, identifiable" (Recital 26, Regulation (EU) 2016/679).
Anonymisation is usually intended as a process through which personal data are manipulated (concealed or hidden) to make it difficult to identify data subjects [8]. This can be done either by deleting or omitting identifying details or aggregating information [9].

However, in the last years the practice has revealed that anonymized data can be often easily re-identified or de-anonymized, especially considering that the wider and

---

[12] Recital 26 of the Regulation (EU) 2016/679 defines "anonymous data" as "the information which does not relate to an identified or identifiable natural person".

wider scale of electronic processing of data and the more and new sophisticated data mining processes and data analytics make possible to combine data sets from so many different sources to derive new information that can lead to re-identification of a person. [8][13]

The issue on risks of re-identification has been recently addressed by the Article 29 Working Party [10] according to which information are not "personal data" only when it is anonymized to the effect that it is no longer possible to associate it to an individual by using "all the means likely reasonably to be used" either by the controller or a third party.

In this opinion of the Article 29 Working Party, the outcome of such kind of anonymisation should be, in the current state of technology, as permanent as erasure, i.e. making it impossible to process personal data. Only in presence of an irreversible anonymisation it is possible to state that the EU Data Protection Law no longer applies. By setting the risk threshold at zero for any potential recipients of the data the consequence is that, except for very limited cases, there are no existing techniques that can achieve the required degree of anonymization. In which case sharing the data requires the consent of the data subjects or, alternatively, to rely on any of the other legitimate grounds provided by the EU Data Protection law.

However, part of the doctrine raised criticism on this strict approach considering that the Article 29 Working Part follows an absolute definition of acceptable risk in the form of "zero risk" while the legislation on the protection of personal data (i.e. the current Directive 95/46/EC on the protection of personal data, and the Regulation (EU) 2016/679) itself does not require a zero risk approach [11]. Indeed, the EU Data Protection Law provides that the impossibility of re-identification should be approached in light of the "all means likely reasonably" test. In this direction it moves indeed the most recent legislation: according to the new wording of Recital 26 of the Regulation (EU) 2016/679 "to determine whether a natural person is identifiable, account should be taken of all the means reasonably likely to be used, such as singling out,[14] either by the controller or by another person to identify the natural person directly or indirectly. To ascertain whether means are reasonably likely to be used to identify the natural person, account should be taken of all objective factors, such as the costs of and the amount of time required for identification, taking into

---

[13] See the AOL and Netflix cases in Ohm Paul, *Id* (2009), 1717-1722.

[14] This is a new wording introduced by the Regulation (EU) 2016/679 and draws on the Opinion of the Article 29 Working Party where it stated that an effective anonymisation solution prevents all parties from singling out an individual in a dataset, from linking two records within a dataset (or between two separate datasets) and from inferring any information in such dataset. Another factor pointed out by the Article 29 Working Party in its Opinion 5/2014 in assessing the notion of "impossibility" is the robustness of the anonymisation technique employed. In assessing the robustness of different techniques of anonymization, the following questions should be taken into account: (1) is it still possible to single out an individual; (2) is it still possible to link records relating to an individual, and (3) can information be inferred concerning an individual? Using these three questions, the Article 29 Working Party produced a table that shows the strengths and weakness of the different techniques of anonymisation.

consideration the available technology at the time of the processing and technological developments". It seems reasonable, indeed, that the wording used by the EU legislator was not intended to adopt a zero risks expectation.

Additionally it would be worth to outline that anonymisation triggers by itself a further processing of personal data and, as such, it must satisfy the requirement of compatibility of the anonymization with the original purposes of the collection. According to the Opinion of Article 29 Working Party [10], for the anonymisation to be considered as compatible with original purposes of the processing, the anonymisation process should produce reliably anonymized information.

In the light of the above, businesses should bear in mind that implementing anonymisation techniques does not exempt them from obligations provided under the EU Data Protection Law. Indeed, the purpose limitation principle still applies even when the further processing is aimed at anonymizing personal data.

Given this, it is recommended for businesses – also where they are not strictly required under mandatory laws[15] - (i) to carry out an effective data protection impact assessment to verify the compatibility of the anonymization with the purposes for which data were originally collected; (ii) to identify what data may be available for sharing and at what level of anonymization and aggregation; and (iii) to detect any risks of re-identification by taking into account also the technological, economic and organizational capacity of the third parties recipients.

Pseudonymization. An alternative, but significantly different method to reduce the possibility to immediate identification of data subjects is the pseudonymization, i.e. the procedure to replace identification data (e.g. the name of an individual or other direct identifiers) with codes or numbers.

According to the Opinion of Article 29 Working Party [10] and the EU Data Protection Law (Recital 28 and Article 4, paragraph 1, n. 5, Regulation (EU) 2016/679), pseudonymization method is a measure to protect personal data, to which extent it can reduce (but does not exclude) the risk related to the processing of personal data. Implementing pseudonymization techniques is indeed very often expressly recommended also by the data protection regulators to help controllers and processors to meet their data protection obligations. In particular, under the EU Data Protection Law pseudonymization is explicitly defined as "the processing of personal data in such a manner that the personal

---

[15] According to Article 35, paragraph 1, of the Regulation (EU) 2016/679 carrying out an assessment of the impact of the processing operations on the protection of personal data is required when, given the type of processing, in particular if new technologies are involved, and taking into account the nature, scope, context and purposes of the processing, it is likely that the processing results in a high risk to the rights and freedoms of natural persons. Pursuant to Article 35, paragraph 3.a, it is specifically required to carry out a data protection impact assessment in case of: (a) systematic and extensive evaluation of personal aspects relating to natural persons which is based on automated processing, including profiling, and on which decisions are based that produce legal effects concerning the natural person or similarly significantly affect the natural person; (b) processing on a large scale of special categories of data or of personal data relating to criminal convictions and offences; or (c) systematic monitoring of a publicly accessible area on a large scale.

data can no longer be attributed to a specific data subject without the use of additional information, provided that such additional information is kept separately and is subject to technical and organizational measures to ensure that the personal data are not attribute to an identified or identifiable natural personal".

The Article 29 Working Party stressed that equating pseudonymized data to anonymized data is considered as one of the misconceptions among many data controllers. Pseudonymized data, indeed, continues to allow an individual to be singled out and linkable across different data sets and their use remains subject to data protection rules. Accordingly, applying pseudonymization technique should not be seen by businesses sharing personal data as a way to avoid application of data protection rules. On the contrary, it should be seen as a measure to strengthen the individuals' data protection rights.

3.2   Sharing Personal Data

If sharing of personal data (as opposed to anonymous information) is strictly necessary to achieve the aim pursued when the data are disclosed, businesses should take the necessary safeguards in order to comply with data protection rules, but also to ensure that third parties receiving personal data will do the same: the disclosing party should honor its obligations towards the individuals from whom it collected personal data and should take the appropriate actions to ensure that the receiving party is bound to certain conditions for the processing of the shared personal data.

Preserving data protection while sharing personal data could be achieved by implementing privacy by design strategies in different phases of the data life-cycle: from the collection to the moment of the re-use by the recipients. The aim of these strategies is indeed to mitigate the risks of unlawful processing by:

– increasing awareness of data subjects about any further processing operations: the more transparency on the data life-cycle controllers guarantee the more likely it is that any further use they contemplate may be considered "compatible" (the compatibility is not required if a new consent is obtained for sharing). Using standardized machine-readable icons, as recommended by the EU Data Protection Law (Recital (60), Regulation (EU) 2016/679) will be one of the preferable manners to offer clear and easily visible transparency to the data subjects [12–14];
– even when this is not strictly mandatory, carrying out a data protection impact assessment in order to identify the risks which may originate from sharing of data as well as to detect possible liabilities for each of the parties involved in the data life-cycle;
– developing appropriate and non–traditional privacy policies which should automatically be enforced against any party involved in the data life-cycle, e.g. using data sharing agreements [15];
– complying with the minimization principle by reducing the identifiability of individual as much as possible, keeping in mind that using pseudonymization techniques does not exclude the risks of re-identification;
– implementing new tools to allow data subjects to control how their data are processed and shared, and to actively handle their rights at any time; this is of a paramount

importance for future since the EU Data Protection Law provides a wide range of rights in favor of data subjects – from the right of erasure (i.e. the right to be forgotten) to the right of data portability –exercise of which by a large group of individuals might undermine the resources of businesses in term of personnel, time and costs.

On the other hand, controllers should also identify implications of relying on a legal ground instead of another.

The EU Data Protection Law, indeed, allows the processing of personal data only to the extent that it relies on, at least, one of the specific and limited legal grounds provided under Article 6.[16] When businesses decide to share personal data on the basis of the data subjects' consent, they may rely on new mechanisms of consent; the traditional notice and consent paradigm gave, indeed, only an apparent, but inconsistent, self-determination [16]. However, over the last years, new technologies have reinforced the concept of consent by creating user friendly consent mechanisms which are mainly based on engineered banner solutions. Moreover, taking into account that the EU Data Protection Law provides that consent can be also given by a statement or by a clear affirmative action, other types of consent tools may be developed which will involve practical user positive actions through new sensors (e.g. gesture, spatial patterns, behavioral patterns, motions [17]). Provided that the EU Data Protection Law (Article 7, paragraph 3, Regulation (EU) 2016/679) requires that it shall be as easy to withdraw consent as to give it, businesses shall also implement appropriate mechanism to allow individuals to easily revoke their consent.

This should be implemented by adopting the approach of privacy preferences according to which sticky policies can provide a mechanism for attaching privacy preferences to specific data sets and accordingly drive data processing decisions.

On the other hand, when businesses should decide to rely on their legitimate interest in sharing data, they shall implement appropriate mechanism in order to inform individuals about the further processing and permit them to exercise their data protection rights, including the right to object, i.e. to opt-out (Article 21, Regulation (EU) 2016/679).

4   Conclusion

Re-use of personal data to share them for purposes other than those for which data were originally collected is not forbidden by the EU Data Protection Law, but, on the contrary,

---

[16] According to Article 5 of the Regulation (EU) 2016/679, personal data can be processed if the data subject has given his/her consent or if the processing is necessary:

(a) for the performance of a contract to which the data subject is party or in order to take steps at the request of the data subject prior to entering into a contract;
(b) for compliance with a legal obligation to which the controller is subject;
(c) in order to protect the vital interests of the data subject or of another natural person;
(d) for the performance of a task carried out in the public interest or in the exercise of official authority vested in the controller; or
(e) for the purposes of the legitimate interests pursued by the controller or by a third party.

it can be carried out on the basis of various legal grounds, e.g. the "compatibility test" or consent.

However, if the sharing of the personal data involves large amount of personal data, it is of paramount importance for businesses to carry out a prior data protection impact assessment in order to identify the roles and the potential risks related to the sharing, and to design strategies for preserving privacy during the entire data chain.

To this extent, the current market scenario has started to provide the first examples on privacy preserving solutions and the technologies bode well for the future.

References

1. European Parliamentary Research Service, The Cost of Non-Europe in the Sharing Economy (2016)
2. European Commission, European Cloud Initiative - Building a competitive data and knowledge economy in Europe – COM (2016). 178 final
3. European Union Agency for Fundamental Rights, Handbook on European Data Protection Law (2014)
4. European Commission. http://europa.eu/rapid/press-release_IP-16-1407_en.htm
5. Information Commissioner's Office, The Data Sharing Code of Practice (2011). https://ico.org.uk/media/for-organisations/documents/1068/data_sharing_code_of_practice.pdf
6. Article 29 Data Protection Working Party, Opinion 03/2013 on Purpose Limitation (WP203), 2 April 2013
7. Fisk, G., Ardi, C., Pickett, N., Heidemann, J., Fisk, M., Papadopoulos, C.: Privacy principles for sharing cyber security data. In: Security and Privacy Workshops (SPW). IEEE (2015)
8. Paul, O.: Broken promises of privacy: responding to the surprising failure of anonymisation. UCLA Rev. **57**, 1707 (2009)
9. Kuan, H.W., Millard, C., Walden, I.: The problem of 'personal data' in cloud computing – what information is regulated? The cloud of unknowing. Int. Data Priv. Law **1**(4), 211–228 (2011). Queen Mary School of Law Legal Studies Research Paper No. 75/2011
10. Article 29 Working Party, Opinion 05/2014 on Anonymisation Technique (WP216), 10 April 2014
11. El Emam, K., Alvarez, C.: A critical appraisal of the article 29 working party opinion 05/2014 on data anonymization techniques. Int. Data Priv. Law **5**(1), 73–87 (2015)
12. Edwards, L., Abel, W.: The Use of Privacy Icons and Standard Contract Terms for Generating Consumer Trust and Confidence in Digital Services, CREATe Working Paper 2014/15, 31 October 2014
13. Holtz, L.-E., Nocun, K., Hansen, M.: Towards displaying privacy information with icons. In: Fischer-Hübner, S., Duquenoy, P., Hansen, M., Leenes, R., Zhang, G. (eds.) Privacy and Identity Management for Life. IFIP AICT, vol. 352, pp. 338–348. Springer, Heidelberg (2011)
14. Enisa, On the security, privacy and usability of online seals. An overview (2013)
15. Caimi, C., Gambardella, C., Manea, M., Petrocchi, M., Stella, D.: Legal and technical perspectives in data sharing agreements definition. In: Berendt, B., et al. (eds.) APF 2015. LNCS, vol. 9484, pp. 178–192. Springer, Heidelberg (2016). doi:10.1007/978-3-319-31456-3_10
16. Mantelero, A.: The future of consumer data protection in the E.U. rethinking the "notice and consent" paradigm in the new era of predictive analytics. Comput. Law Secur. Rev. **30**(6), 643–660 (2014)
17. Enisa, Privacy by Design in Big Data (2015)

2nd International Workshop on Mining the Social Web (SoWeMine 2016)

Identifying Great Teachers Through Their Online Presence

Evanthia Faliagka[1(✉)], Maria Rigou[2,3], and Spiros Sirmakessis[1,3]

[1] Department of Computer and Informatics Engineering,
Technological Institution of Western Greece, Antirrio, Greece
{efaliaga,syrma}@teiwest.gr
[2] Department of Computer Engineering and Informatics, University of Patras,
Rion Campus, Patras, Greece
rigou@ceid.upatras.gr
[3] Hellenic Open University, Patras, Greece

Abstract. Evaluating candidate teachers is a very tricky task, as there are a lot of criteria -objective and not- that are important for identifying a good teacher. The teacher's efficiency depends on the academic qualifications and experience, on teacher's personality, even the students of the class and how well teaching and learning dynamically 'grows'. In this work we propose a novel approach for teacher online evaluation. We implemented a prototype system which extracts values for a set of objective criteria from the teachers' LinkedIn profile, and infers personality characteristics using linguistic analysis on their Facebook and Twitter posts. Machine learning algorithms were used to solve the final ranking problem.

Keywords: E-recruitment systems · Personality mining · Personality traits · Social web mining · Recommendation systems · Teacher evaluation

1 Introduction

Successful teachers are not just those who are well-educated and possess the required knowledge to communicate to students. Teachers are also judged on the basis of their personality, as personality is considered one of the most important factors both for teaching, as well as for learning and academic achievement and it is a common belief that teachers with certain personality traits are able to teach more efficiently [1, 2]. Numerous studies have examined which personality characteristics have a positive or a negative effect in the performance of a teacher and distinguishing him/her as 'liked' (excellent, effective, good, qualified) or 'disliked' (hated, amateur and inefficient) [3–7]. According to studies, teachers are characterized as liked or disliked based on three criteria: academic qualifications, relationship with students, and personality traits [4, 6, 8–10]. As far as teacher personality is concerned, teachers should be humble [4, 11], polite and friendly [12], serious, eager to teach, fond of their job [8, 12], warm, cheerful, and well-balanced [13]. In addition, teachers should have creative and flexible viewpoints and high levels of cognitive proficiency and creativity [4, 7, 14, 15]. Furthermore, some studies on the characteristics put emphasis on conscientiousness [8, 12],

© Springer International Publishing AG 2016
S. Casteleyn et al. (Eds.): ICWE 2016 Workshops, LNCS 9881, pp. 71–79, 2016.
DOI: 10.1007/978-3-319-46963-8_6

agreeableness [4, 13], openness to experience, and extroverted personality traits in order to yield positive educational results [4, 7, 14]. In the relevant literature, the approach which includes all the aforementioned features is the Big-Five Personality Model [16]. This personality model is adapted in the present approach and is presented in Sect. 4.

In the relative literature there are tests available (in the form of questionnaires) that evaluate the personality characteristics of an individual appropriately designed on the basis of the respective theoretical personality model. People responsible for recruiting personnel in big firms or for hiring people in job positions that pose by nature specific personality prerequisites use either special purpose questionnaires/tests or interviews to evaluate the adequacy of a candidate for a certain position.

Nowadays though, the amount of information available at all levels of people's social environment has drastically increased [17] and we have been steadily turning to the web for social interaction and recreation, but also in order to improve our skills [18], for career development [19], as well as for searching for a job. Job seekers are increasingly using web 2.0 services like LinkedIn and job search sites [20], while a lot of companies use online knowledge management systems to hire employees, exploiting the advantages of the World Wide Web. These are termed e-recruitment systems and automate the process of publishing positions and receiving CVs. The online recruitment problem is two-sided: It can be seeker-oriented or company-oriented. In the first case the e-recruitment system recommends to the candidate a list of job positions that better fit his profile. In the second case recruiters publish the specifications of available job positions and the candidates can apply.

In on-line recruitment systems, candidates typically upload their CVs in the form of a document with a loose structure, which must be considered by an expert recruiter. However this incorporates a great asymmetry of resources required from candidates and recruiters and potentially increases the number of unqualified applicants. This situation might be overwhelming to HR agencies that need to allocate human resources for manually assessing the candidate resumes and evaluating the applicants' suitability for the positions at hand. Several e-recruitment systems have been proposed with an objective to automate and speed-up the recruitment process, leading to a better overall user experience and increasing efficiency. For example, SAT telecom reported 44 % cost savings and a drop in the average time needed to fill a vacancy from 70 to 37 days [21] after deploying an e-recruitment system.

When recruiting concerns teacher positions the process could be performed online provided that we can have some unbiased feedback on the personality of the candidate teacher. Social web data (coming from Facebook, Twitter, Linkedin, etc.) can be the source of such feedback especially in the case of active users, if we can interpret social web activities in personality traits demonstrated.

In this work, we have implemented an integrated system that automates the teachers' evaluation process. Its objective is to calculate the teachers' relevance scores, which reflect how well their profile fits the positions' specifications. The system is a variation of the e-recruitment system presented in [22–24] customized for assessing the desired personality traits of teachers. The next sections provide details regarding the architecture of the proposed system comprising three main modules, followed by the description of the personality module. The discussion moves to the pilot recruitment scenario that was used to evaluate the effectiveness of the proposed approach and the paper ends with the main conclusions reached.

2 Architecture

The proposed system implements automated candidate ranking based on a set of credible criteria. In this study we focus on 4 complementary selection criteria, namely: Education (in years of formal academic training), Work Experience, Skills and Personality. The system architecture, which is depicted in Fig. 1, consists of the following components:

- *Job Application module:* It implements the input forms that allow teachers to apply for a class. The teacher is given the option to log into the system using the personal LinkedIn account credentials, thus allowing the system to automatically extract all objective selection criteria directly from the user's LinkedIn profile.
- *Personality mining module*: If the candidate's twitter URL is provided, the system applies linguistic analysis to tweets to derive features reflecting the author's personality. The linguistic analysis is done with the LIWC tool that uses a dictionary of word stems classified in certain psycholinguistic semantic and syntactic word categories [25]. The teacher can also sign in using the personal Facebook account and the system uses Facebook posts in the linguistic analysis in addition to the tweets.
- *Applicant Grading module*: It combines the candidate's selection criteria to derive the candidate's relevance score for the applied position. The grading function is derived through supervised learning algorithms.

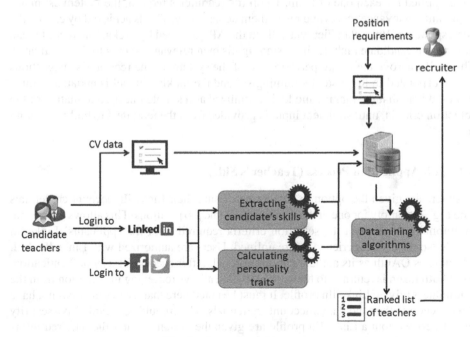

Fig. 1. System's architecture.

The system is to be used as a filtering and assistive tool in making successful hiring decisions. Each teacher's qualifications, as well as his relevance score are stored in the system's database. At the end of the recruitment process, the top n teachers could be called to participate in a trivial interview process. It must be noted here that during the job application process, the applicant is not required to manually enter information or participate in time-consuming personality tests. Thus, the user friendliness and the practicality of the system are maintained.

3   Job Application Module

The proposed system was fully implemented as a web application, in the Microsoft.Net development environment. In this section we will present indicative application screens and discuss our design decisions and system implementation. The system is divided in the recruiter's side and the user's side.

3.1   Recruitment Process (Recruiter's Side)

After authenticating with their account credentials, recruiters have access to the recruitment module, which gives them rights to post new job positions and evaluate job applicants. For each new job position a recruiter can define the prerequisites for the position and the skills that are required for the specific class. In the "rank teachers" menu, the recruiter is presented with a list of all available job positions and the candidates that have applied for each one of them. Upon the recruiter's request, the system estimates applicants' relevance scores and ranks them accordingly. This is achieved by calling the corresponding Weka classifier, via calls to the API provided by Weka. The recruiter can modify the candidate ranking, by assigning his own relevance scores to the candidates. This will improve the future performance of the system, as the recruiter's suggestions are incorporated in the system's training set and the ranking model is updated. It must be noted here that the ranking model is initialized as a simple linear combination of the selection criteria, until sufficient input is provided from the recruiters to build a training set.

3.2   Job Application Process (Teacher's Side)

Teachers are given the option to authenticate using their LinkedIn account credentials (see Fig. 2) to apply for one or more of the available job positions. This allows the system to automatically extract the selection criteria required from the applicants' LinkedIn profile, so the user experience is streamlined. Users are authorized with LinkedIn API, which uses OAuth as its authentication protocol. After successful user authentication, an OAuth token is returned to the system, which allows retrieving information from the candidate's private LinkedIn profile. It must be noted here that the system does not have direct access to the candidate's account credentials, which could be regarded as a security risk. Users without a LinkedIn profile are given the option to enter the required information manually.

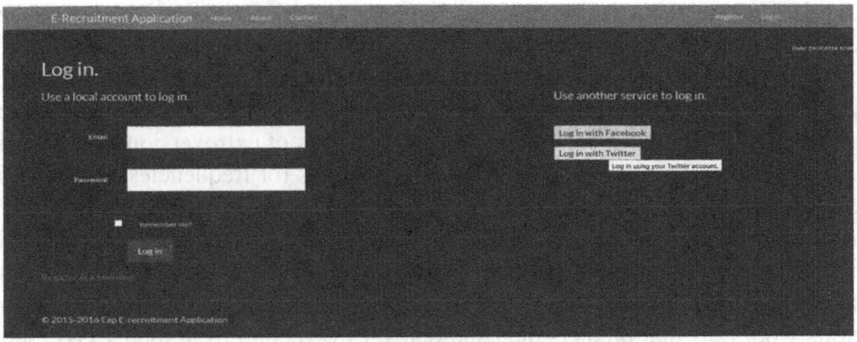

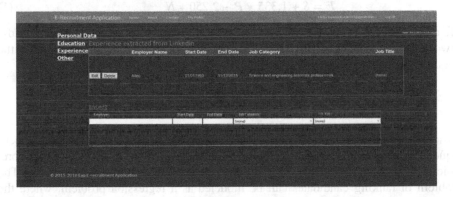

Fig. 2. Login page.

As part of the job application process, the candidate is asked to login with his twitter credentials. This allows our system to syndicate the twitter content and calculate the personality score with the personality mining technique presented below (Fig. 3).

Fig. 3. Experience mining.

4 Extracting Teacher's Personality

The Big-Five Personality Model is the most commonly used in academic psychology, and is also known as NEO PI-R. It measures the personality traits on five domains namely: neuroticism, extraversion, openness to experience, agreeableness and conscientiousness. Studies have examined the relation between its domains and students' evaluation of teaching. One such study [26] that was based on teacher evaluation done by their students and aimed at investigating liked, disliked and neutral teachers, concluded that the most important personality traits of "liked" teachers are extroversion, conscientiousness, agreeableness, emotional stability, and openness. "Disliked" teachers have such personality traits as introversion, suspiciousness and antagonism towards others, emotional instability, an easy-going nature/carelessness and consistency/cautiousness.

In this work we focus on the extroversion personality trait, due to its importance in identifying "liked" teachers. Moreover, it has been shown that extroversion is sufficiently reflected in written speech and can be identified through text analysis. Specifically, the emotional positivity and social orientation of candidates, both directly extracted from LIWC frequencies, can act as predictors of extroversion trait. We estimate the extraversion score directly from LIWC scores (or frequencies), by summing the emotional positivity score and the social orientation score, also obtained from LIWC frequencies, as we proposed in [23]. The main difference is that we no longer use blog posts but Facebook posts and tweets as they currently are concerned the most popular social applications. Finally, we used the regression model which was trained in a previous work [23] that predicts the candidates' extraversion from their LIWC scores in the {posemo, negemo, social} categories. The regression model selected as a predictor of the extraversion score was proposed in [27], due to its good accuracy and low complexity. Equation (1) corresponds to the linear model that minimizes the Mean Square Error between actual values assigned by the recruiter and predicted scores output by the model:

$$E = S + 1.335 * P - 2.250 * N \tag{1}$$

where E is the extraversion score, S the frequency of social words (such as friend, buddy, coworker) returned from LIWC, P the frequency of positive emotion works and N the frequency of negative emotion words.

5 Final Ranking

To solve the problem of ranking candidate teachers we employ machine learning techniques. In the ranking problem, a scoring function $h(x)$ outputs the relevance score, which reflects how well a teacher's profile fits the requirements of a given position. The problem of ranking candidates can be modeled as a regression problem, where the scoring function is learned with supervised learning techniques. Then the ranked list of candidates is derived by applying the learned function to the original list of candidates. The scoring function $h(x)$ derives the teacher's relevance degree y_i from the values of his feature vector x_i. The feature vector x_i consists of a set of m attributes $\{a_1, ..., a_m\}$ that correspond to the candidate's selection criteria. These can be either continuous variables (representing a feature assessed on numerical scale) or Boolean variables (i.e., whether a candidate has a desired skill or not). The actual scoring function is typically unknown and an approximation is derived from the training set D. In the proposed system the training set consists of a set of N previous candidate selection examples, given as an input to the system:

$$D = \left\{ (x_i, y_i) \mid x_i \in R^m, y_i \in R \right\}_{i=1}^{N} \tag{2}$$

6 Pilot Scenario

Our system was tested with in a real-world pilot scenario. Specifically, candidate teachers for a job position offered by a private elementary school agreed to participate in a scenario that we set up specifically to test the reliability of our system. The teachers logged in the system with their LinkedIn credentials and applied for a job position that was also announced through the system. The objective criteria were calculated using the LinkedIn data and then the teachers were connected to their Facebook and Twitter account, so as to extract and calculate their extroversion scores.

The same teachers were also evaluated manually by the school administrator, who evaluated the academic qualifications and assessed their extroversion using face-to-face interviews in a grading scale of 0–5. The automated scores were compared to the manual scores using Weka [28] and the learning-to-rank models were evaluated. Specifically, we used Weka to test the correlation of the scores output from the system (i.e. model predictions) with the actual scores assigned by the recruiters, using the Pearson's correlation coefficient metric.

The performance of the proposed system was evaluated based on how effective it is in discriminating the top k candidates providing a rank that is consistent with the one provided by the human recruiters. Three metrics were used for comparing rankings; the simplest one is the overlap size of the top-k list selected by the system and the human recruiter for each job position, where k = 8 corresponds to 20 % of overall applicants. The second metric is the correlation coefficient (Spearman's rho) of the top-k candidates per category. The third metric is the mean absolute difference (ranking error) of top-k candidate's ranks. It can be seen in Table 1 that the system's scores are accurate enough since it was able to achieve a correlation coefficient of up to 0.72, it outputs a top-8 list that overlapped 75 % and the ranking error reached 2.6.

Table 1. Performance evaluation metrics.

K = 8	Top-k	Correlation	Ranking error
Candidates	6 (75 %)	0.72	2,6

7 Conclusions

In this paper we have presented a novel approach for automatically ranking candidate teachers. The proposed scheme relies on objective criteria extracted from the applicants' LinkedIn profiles and subjective criteria extracted from their social presence, to estimate applicants' overall relevance scores and infer their personality traits. Candidate ranking is based on machine learning algorithms that learn the scoring function based on training data provided by a human operator (i.e. the school administrator). An integrated system was implemented based on the proposed scheme. We validated the system with real-world data.

For the future we plan to investigate potentially better fine-tuning in the algorithm that assesses the personality scores, and set up recruitment scenarios in various domains. The extroversion score could also be assessed using additional social network metrics such as LinkedIn endorsements and recommendations, number of re-tweets, Facebook likes and shares, etc. It would also be interesting for the training of the system, instead of assessing manually the teachers to use special purpose personality assessment questionnaires and also to explore possible mechanisms for assessing teachers' scores in agreeableness and openness to experience.

References

1. De Raad, B., Schouwenburg, H.: Personality in learning and education: a review. Eur. J. Pers. **10**, 303–336 (1996)
2. Entwistle, N., Entwistle, D.: The relationships between personality, study methods and academic performance. Br. J. Educ. Psychol. **40**, 132–143 (1970)
3. Amon, S., Reichel, N.: Who is the ideal teacher? Am I? Similarity and difference in perception of students of education regarding the qualities of a good teacher and of their own qualities as teachers. Teach. Teach.: Theory Pract. **13**(5), 441–464 (2007)
4. Goldstein, G., Benassi, V.: Students' and instructors beliefs about excellent lecturers and discussion leaders. Res. High. Educ. **47**(6), 685–707 (2006)
5. Grieve, A.M.: Exploring the characteristics of "teachers for excellence:" teachers' own perceptions. Eur. J. Teach. Educ. **33**(3), 265–277 (2010)
6. Montalvo, G., Mansfield, E., Miller, R.: Liking or disliking the teacher: student motivation, engagement and achievement. Eval. Res. Educ. **20**(3), 144–158 (2007)
7. Polk, J.A.: Traits of effective teachers. Arts Educ. Policy Rev. **107**(4), 23–29 (2006)
8. Beishuizen, J.J., Hof, E., Putten, C.M., van Bouwmeester, S., Asscher, J.J.: Students' and teachers' cognitions about good thinking. Br. J. Educ. Psychol. **71**, 185–201 (2001)
9. Hill, J.S., Christian, T.Y.: College student perceptions and ideals of teaching: an exploratory pilot study. Coll. Stud. J. **46**(3), 589–601 (2012)
10. Kyriacou, C., Stephens, P.: Student teachers' concerns during teaching practice. Eval. Res. Educ. **13**(1), 18–31 (1999)
11. Thibodeau, G.P., Hillman, S.J.: In retrospect: teachers who made a difference from the perspective of pre-service and experienced teachers. Education **124**(1), 168–181 (2003)
12. Bennett, S.K.: Student perceptions of and expectations for male and female instructors: evidence relating to the question of gender bias in teaching evaluation. J. Educ. Psychol. **74**, 170–179 (1982)
13. Larsgaard, J.O., Charles, E., Kelso, J.R., Thomas, W., Schumacher, M.S.: Personality characteristics of teachers serving in Washington State Correctional Institutions. J. Correctional Educ. **49**(1), 20–38 (1998)
14. Eilam, B., Vidergor, H.E.: Gifted Israeli students' perceptions of teachers' desired characteristics: a case of cultural orientation. Roeper Rev. **33**, 86–96 (2011)
15. Erdle, S., Murray, H.G., Rushton, J.P.: Personality, classroom behavior, and student ratings of college teaching effectiveness: a path analysis. J. Educ. Psychol. **77**, 394–407 (1985)
16. McCrae, R.R., Costa, P.T.: Personality in Adulthood. The Guilford Press, New York (2003)
17. Neuman, C.: Prospero: a tool for organizing internet resources. Internet Res. **20**, 408–419 (2010)
18. Ho, L., Kuo, T., Lin, B.: Influence of online learning skills in cyberspace. Internet Res. **20**, 55–71 (2010)
19. Jansen, B., Jansen, K., Spink, A.: Using the web to look for work: implications for online job seeking and recruiting. Internet Res. **15**, 49–66 (2005)

20. Bizer, R.H., Rainer, E.: Impact of Semantic web on the job recruitment Process, Wirtschaftsinformatik. Physica-Verlag HD (2005)
21. Pande, S.: E-recruitment creates order out of chaos at SAT telecom: system cuts costs and improves efficiency. Hum. Resour. Manag. Int. Dig. **19**, 21–23 (2011)
22. Faliagka, E., Tsakalidis, A., Tzimas, G.: An integrated e-recruitment system for automated personality mining and applicant ranking. Internet Res. **22**(5), 551–568 (2012)
23. Faliagka, E., Iliadis, L., Karydis, I., Rigou, M., Sioutas, S., Tsakalidis, A., Tzimas, G.: Online consistent ranking on e-recruitment: seeking the truth behind a well-formed CV. Artif. Intell. Rev. **42**, 515–528 (2014)
24. Faliagka, E., Rigou, M., Sirmakessis, S.: An e-recruitment system exploiting candidates' social presence. In: Daniel, F., Diaz, O. (eds.). LNCS, vol. 9396, pp. 153–162. Springer, Heidelberg (2015). doi:10.1007/978-3-319-24800-4_13
25. Pennebaker, J.W., King, L.A.: Linguistic styles: language use as an individual difference. J. Pers. Soc. Psychol. **77**(6), 1296–1312 (1999)
26. Eryilmaz, A.: Perceived personality traits and types of teachers and their relationship to the subjective well-being and academic achievements of adolescents. Educ. Sci.: Theory Pract. **14**(6), 2049–2062 (2014)
27. Mairesse, F., Walker, M., Mehl, M., Moore, R.: Using linguistic cues for the automatic recognition of personality in conversation and text. J. Artif. Intell. Res. **30**, 457–500 (2007)
28. Hall, M., Frank, E., Holmes, G., Pfahringer, B., Reutemann, P., Witten, I.H.: The WEKA data mining software: an update. ACM SIGKDD Explor. Newslett. **11**(1), 10–18 (2009)

Tracking Dengue Epidemics Using Twitter Content Classification and Topic Modelling

Paolo Missier[1(✉)], Alexander Romanovsky[1], Tudor Miu[1], Atinder Pal[1],
Michael Daniilakis[1], Alessandro Garcia[2], Diego Cedrim[2],
and Leonardo da Silva Sousa[2]

[1] School of Computing Science, Newcastle University,
Newcastle upon Tyne, UK
paolo.missier@newcastle.ac.uk
[2] PUC-Rio, Rio de Janeiro, Brazil

Abstract. Detecting and preventing outbreaks of mosquito-borne diseases such as Dengue and Zika in Brasil and other tropical regions has long been a priority for governments in affected areas. Streaming social media content, such as Twitter, is increasingly being used for health vigilance applications such as flu detection. However, previous work has not addressed the complexity of drastic seasonal changes on Twitter content across multiple epidemic outbreaks. In order to address this gap, this paper contrasts two complementary approaches to detecting Twitter content that is relevant for Dengue outbreak detection, namely supervised classification and unsupervised clustering using topic modelling. Each approach has benefits and shortcomings. Our classifier achieves a prediction accuracy of about 80 % based on a small training set of about 1,000 instances, but the need for manual annotation makes it hard to track seasonal changes in the nature of the epidemics, such as the emergence of new types of virus in certain geographical locations. In contrast, LDA-based topic modelling scales well, generating cohesive and well-separated clusters from larger samples. While clusters can be easily re-generated following changes in epidemics, however, this approach makes it hard to clearly segregate relevant tweets into well-defined clusters.

1 Introduction

Mosquito-borne disease epidemics are increasingly becoming more frequent and diverse around the globe and it is likely that this is only the early stage of epidemic waves that will continue for several decades. Rapidly spreading diseases to combat nowadays are those transmitted by the *Aedes* mosquitoes [CDC15], which carry not only *Dengue* virus, but also *Chikungunya* and *Zika* viruses [CDC15], which are responsible for thousands of deaths every year. Therefore, improved surveillance through rapid response measures against Aedes-borne diseases is a long-standing tenet to various health systems around the world. They are urgently required to mitigate the already heavy burden on those health systems and limiting further spread of mosquito-borne diseases within geographical locations, such as in Brazil. Control of Aedes-borne disease requires the vector control by identifying and reducing breeding sites.

© Springer International Publishing AG 2016
S. Casteleyn et al. (Eds.): ICWE 2016 Workshops, LNCS 9881, pp. 80–92, 2016.
DOI: 10.1007/978-3-319-46963-8_7

Our approach to addressing this problem involves the automatic detection of relevant content in Twitter, in order to determine its relevance as actionable information. Paraphrasing [SOM10], we note that social media users are increasingly viewed as *informative social sensors*, who spontaneously communicate valuable information, which in this case may help in detecting the location and extent of mosquito foci. However, as the signal produced by these sensors is very noisy, our realistic goal is to automatically categorise Twitter messages into a few classes, segregating recognisable highly informative from less informative and noisy content.

Previous work, e.g. [GVM+11, LC10, ALG+11], has identified the potential of social media channels, such as Twitter, on offering continuous source of epidemic information, arming public health systems with the ability to perform real-time surveillance. However, previously proposed approaches are often limited or insufficient for rapid combat of epidemic waves for several reasons. Firstly, previous work has mainly explored the use of social media channels to predict Dengue cases and outbreak patterns by exploring disease-related posts from previous outbreaks. However, the combination of socio-economic, environmental and ecological factors dramatically changes the characteristics governing each epidemic wave. As a consequence, exploring disease-related posts from previous outbreaks tends to be ineffective to identify breeding sites in the outset of each outbreak. Secondly, previous work is not aimed at identifying map breeding sites of the mosquito within a region.

The role of Twitter content relevance detection is depicted in Fig. 1. Social sensors, the people in the upper half of the figure, contribute information either

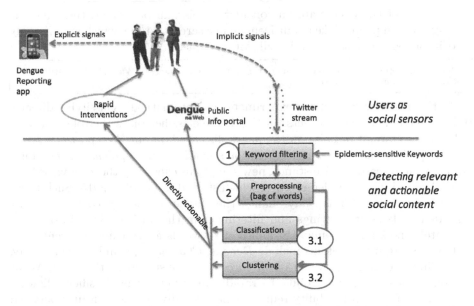

Fig. 1. Role of automated Twitter relevance detection for health vigilance against Dengue

implicitly, i.e., by spontaneously carrying out public conversations on social media channels, or explicitly, i.e., by interacting with dedicated public Web portals and mobile apps. As an example, our group in Brazil has been developing both such a Dengue mapping portal, and a mobile app that members of the public may use to report cases of Dengue in their local areas [PR15].

As shown in the figure, we monitor the Twitter feed, pre-select tweets according to a broad description of the Dengue topics using keywords, then classify the selected tweets, aiming to segregate relevant signal from the noise. We distinguish the relevant signal further into directly and indirectly actionable information. Directly actionable tweets, which we classify as *mosquito focus*, are those that contain sufficient information regarding a breeding site (including geo-location), to inform immediate interventions by the health authorities. For instance:

> @Ligue1746 Atenção! Foco no mosquito da dengue. Av Sta Cruz, Bangu. Em frente ao hospital São Lourenço! (*@Ligue1746 Attention! Mosquito focus found in Santa Cruz avenue, Bangu. In front of the So Loureno hospital!*)

These posts are relatively scarce within the overall stream, however, accounting for about 16 % of the ground truth class assignments. Indirectly actionable tweets carry more generic information about members of the public complaining about being affected by Dengue (the *Sickness* class), or *News* about the current Dengue epidemics. For example:

> Eu To com dengue (*I have dengue fever*)
> ES tem mais de 21 mil casos de dengue em 2015 (*ES has more than 21 thousands cases of dengue in 2015*)

The rest of the tweets are all considered noise. In particular, these include messages where people joke about Dengue in a sarcastic tone, which is commonly used in online conversation in Brazil, for example:

> Meu WhatsApp ta tão parado que vai criar mosquito da dengue (*My WhatsAp is so still that it'll create dengue mosquito*)

In this paper we report our experiments on automatically classifying directly and indirectly actionable tweets. In Fig. 1, the classifier plays the role of a filter to limit the amount of noise on the pages displayed on our Web portal.

One problem faced in our classification scenario is that *epidemic waves* differ from season to season. For instance, new symptoms caused by the *Zika* virus have been observed in the epidemic wave, which started in October 2015. Such types of epidemic changes drastically change the nature of Twitter content, requiring different keyword settings and filtering from the Twitter feed, in order to accurately track an epidemic. Examples of keywords for tracking different virus and new emerging symptoms include *Dengue*, *Chikungunya*, and, more recently, *Zika*. Simply taking the union of all three would just add to the noise. What is required instead is the ability to rapidly reconfigure the classifier following a drift in topic. This flexibility requirement naturally suggests an unsupervised approach to training the classifier. At the same time, a supervised classifer that is trained using manually labelled content is likely to be more accurate.

1.1   Contributions

In this work we explore the trade-offs between accuracy and flexibility, by comparing and contrasting a supervised classifier learning approach (3.1 in Fig. 1) with an unsupervised content clustering, using Topic Models (3.2) and specifically on LDA [BNJ03], a popular algorithm that has been previously shown to apply well to clustering Twitter data [RDL10, REC12]. We expect supervised classification to provide good accuracy, as well as give an obvious way to select actionable content from the most informative classes (*mosquito focus*, *sickness*, and *News* in this order). On the other hand, this model suffers from known limitations in the size of the training set, which may lead to disappointing performance on content in the wild, and it is expensive to re-train following changes in the filtering keywords.

In contrast, topic modelling is a form of semantic clustering where a clustering scheme can be easily periodically re-generated from large samples. While the clear characterisation of clusters using ranked lists of terms from the content's vocabulary (topics) makes this a popular approach, a topic may include heterogeneous content that cuts across expert-defined classes, such as those above, making it harder to associate them with a clear focus. This problem is particularly acute in our setting, where we already have a topic defined (through keywords), and we are essentially asking LDA to further refine it in terms of well-separated sub-topics.

We assessed the potential of our approach on large cities of Brazil, such as Rio de Janeiro, by analyzing two cycles of Aedes-related epidemic waves. Our specific contributions in this paper are: (i) a pipeline that implements both methods, including a dedicated pre-processing phase that accounts for idiosyncratic use of the Brazilian Portuguese language in tweets, and (ii) an experimental evaluation of their effectiveness. The supervised classifier is currently in operation as part of the experimental Dengue Web portal developed at PUC-Rio [PR15].

1.2   Related Work

This paper makes original contributions to an already existing landscape of research on monitoring social media for health vigilance purposes. Similarly to our work, Twitter data is used by [GVM+11] to track the Twitter stream and filter relevant signals from it. Because they only use supervised classification for content filtering, their approach is limited by the amount of labels made available by expert annotators. Moreover, this limitation does enable to easily reveal new information in the outset of each epidemic wave. In our work, we use not only supervised classification, but also unsupervised clustering as means of identifying relevant social signals. Finally, we contrast the results from both methods in order to: (i) reflect on the different use cases the methods require (i.e. in terms of annotation effort), and (ii) observe how unsupervised classification helps to better achieve the purpose of revealing new information in each epidemic wave.

[ALG+11, LC10] show that the frequency of tweets containing simple search keywords can be a good indicator of a trend for a flu epidemic. The authors show that there is a strong correlation between the number of medically registered visits to a GP concerning flu and the number of tweets mentioning flu. This approach to tracking epidemics is complementary to ours because, while the previously mentioned authors measure tweet activity on an entire corpus of tweets, we use machine learning to further discover sub-signals in the corpus in specific epidemic waves. Our approach enables one to further measure and study tweet activity within relevant sub-signals.

Similar methods of monitoring Twitter data have been applied for general event detection, as done, for example, by [CW14] or [BNG11]. However, obtaining ground truth is recognised to be a serious bottleneck in a supervised learning pipeline and efforts to reduce the annotation effort have been attempted. For instance, in [GBH09] the ground truth from emoticons for sentiment classification is automatically identified. However, as previously mentioned, even if ground truths are somehow identified, the use of supervised learning may not suffice to cope with tracking the changing characteristics of different epidemic waves.

2 Twitter Content Acquisition and Processing

Our experimental dataset consists of three sets of Twitter content, harvested over two periods of time, during the first and second semester of 2015. These periods corresponded to two cycles of epidemic waves. The first two sets, of about 1,000 and 1,600 instances, respectively, were manually annotated by our group at PUC-Rio, which also included the participation of a medical doctor and an epidemiologist. They were used in supervised classification as our training and test set (using standard k-fold validation), and for further testing (no training), respectively. A larger third set of about 100,000 tweets was used for topic modelling.

A technique similar to that described in [NGS+09] was used to determine a set of filtering keywords for harvesting the tweets. Namely, we started with the single #dengue hashtag "seed" for an initial collection. Upon manual inspection of about 250 initial tweets, our local experts then extended the set to include the most relevant hashtags. These hashtags were those that all local experts agreed to be relevant after discussion amongst them. The final search set contains the following elements (including their common minor variations): { #Dengue, #suspeita, #Aedes, #Epidemia, #aegypti, #foco, #governo, #cuidado, #febreChikungunya, #morte, #parado, #todoscontradengue, #aedesaegypti}.[1]

Content pre-processing includes a series of normalisation steps, followed by POS tagging and lemmatisation.[2] We normalised the content by removing 38 kinds of "twitter lingo" abbreviations, some of which are regional to Brazil ("abs"

---

[1] Only tweets in the Portuguese language were considered in this study.

[2] We used the tagger from Apache OpenNLP 1.5 series (http://opennlp.sourceforge.net/models-1.5/), and the LemPORT Lemmatizer customised for Portuguese language vocabulary.

for "abraço", "blz" for "beleza", etc.), as well as all emoticons and non-verbal forms of expressions. While those are crucial to understanding the *sentiment* expressed in a tweet, we found that they are not good class predictors, including the *Jokes* class. We also replaced links, images, numbers, and idiomatic expressions using conventional terms (*url, image, funny,...*).

3 Supervised Classification

Our classification goal has been to achieve a finer granularity of tweet relevance than just a binary classification into actionable and noise. The following set of four classes, of decreasing relevance, gave us at the same time a good accuracy and granularity:

Mosquito-focus: this is the most *directly actionable* class, including tweets that report sites that are or may be foci for Dengue mosquito, or sites that provide conducive environments to mosquito breeding. This class accounts for about 16 % of tweets in our test set.

Sickness: This is the second most informative class. These tweets represent cases of: (i) users suspecting or confirming they are sick or they are aware of somebody else who is sick, and (ii) users discussing disease symptoms. Note that previous work (Sect. 1.7) on tracking Aedes-related epidemic waves make no distinction between this and the previous class.

News: This class represents general news about Dengue, i.e. tweets that spread awareness, report on available preventive measures, inform about health campaigns, and report the number of Dengue cases in certain locations. These are stilll *indirectly actionable* and useful e.g. to show emerging outbreak patterns in specific areas.

Joke: Finally, about 20 % of the tweets in our sample contain a combination of jokes or sarcastic comments about Dengue epidemic. While we regard these as noise, their detection requires an understanding of sarcastic tone in short text, which is challenging as it uses the same general terms as those found in more informative content.

The training set of about 1,000 messages was annotated by three local experts independently, by taking the majority class for each instance, requiring about 100 hours over three refinement steps to resolve inconsistencies and ambiguities. The classes are fairly balanced: *News*: 333 (31 %), *Joke*: 148 (14 %), *Mosquito focus*: 257 (24 %), and *Sickness*: 338 (31 %). Classification performance, measured using standard cross-validation, was similar across different classifier models, namely SVM, Naive Bayes, and MaxEntropy. We chose Naive Bayes as having probabilities associated to each class assignment helped identify the weak assignments, and thus the potential ambiguities in the manual annotations.

The classifier reported an overall 84.4 % accuracy and .83 F-measure. In order to further validate these results, we then sampled an additional set of 1,600 tweets, none of them used for training, and performed both automated classification and manual annotation on this set. On this new set, the distribution of instances in each class, taken from the ground truth annotations, is not substantially different from that in the training set, except for the more abundant *Mosquito focus* class: *News*: 404 (25 %), *Joke*: 289 (18 %), *Mosquito focus*: 253 (16 %), and *Sickness*: 649 (41 %). Performance results for this classifier are reported in Table 1.

Table 1. Classifier performance on independent test set

Class	Precision	Recall	F	Accuracy
News	.79	.74	.76	.74
Joke	.63	.85	.72	.85
Mosquito focus	.79	.85	.83	.86
Sickness	.91	.78	.84	.78

4   Unsupervised Content Clustering Using LDA

As discussed earlier, supervised classification does not fully meet our requirements, as manual annotation limits the size of training set and makes it difficult to update the model when the characteristics of the epidemics changes. Also, finding a crisp, unambiguous classification has been problematic.

LDA-based clustering [BNJ03] has been used before for Twitter content analysis and topic discovery, for example by [MPLC13, DSL+11, WLJH10]. What we investigate is an application of LDA that shows the potential for scalability and flexibility, i.e., by periodically rebuilding the clusters to track drift in Twitter search keywords.

For this experiment, our sample dataset consists of 107, 376 tweets, harvested in summer 2015 using standard keyword filtering from the Twitter feed, and containing a total of 17, 191 unique words. Raw tweets were pre-processed just like for classification (phase 2 in Fig. 1), producing a bag-of-words representation of each tweet. Additionally, as a further curation step we removed the 20 most frequent words in the dataset, as well as all words that do not recur in at least two tweets. This last step is needed to prevent very frequent terms from appearing in all topics, which reduces the effect of our cluster quality metrics and cluster intelligibility.

4.1   Evaluation of Clustering Quality

We explored a space of clustering schemes ranging from 2 to 8 clusters.[3] In the absence of an accepted gold standard, a number of evaluation methods have been proposed in the literature. For instance, [MPLC13] proposes to measure

---

[3] All experiments carried out using the Apache Spark LDA package https://spark. apache.org.

cluster quality by quantifying the differences caused in topic mining using two different stream sampling mechanisms. The method is based on the differences between the distribution of words across topics and between the two sampling mechanisms. However, it cannot be used in our setting, because our corpus of tweets is fixed, rather than a sample. Also, while any two individual words may have different frequency distributions, the approach does not necessarily take into account the importance, measured by relative frequency, of the words within the entire corpus. In an alternative approach, [DSL+11] use ground truth in the form of pre-established hashtags. This is not applicable in our scenario, either, because by the way our topic filtering is done, most of the tweets in our corpus will already include a high number of hashtags, including for instance the #dengue hashtag.

Instead, we propose to use *intra-* and *inter-* cluster similarity as our main evaluation criteria. This is inspired by *silhouettes* [Rou87], and based on the contrast between *tightness* (how similar data are to each other in a cluster) and *separation* (how dissimilar data are across clusters). Specifically, we define the similarity between two clusters C_a, C_b in terms of the cosine TF-IDF similarity of each pair of tweets they contain, i.e., $t_i \in C_a$ and $t_j \in C_b$, as follows:

$$sim(C_a, C_b) = \frac{1}{|C_a|\,|C_b|} \sum_{t_i \in C_a, t_j \in C_b} \frac{\mathbf{v}(t_i) \cdot \mathbf{v}(t_j)}{||\mathbf{v}(t_i)||\,||\mathbf{v}(t_j)||} \qquad (1)$$

where $\mathbf{v}(t_i)$ is the TF-IDF vector representation of a tweet. That is, the kth element of the vector, $t_i[k]$, is the TF-IDF score of the kth term. As a reminder, the TF-IDF score of a term quantifies the relative importance of a term within a corpus of documents [AZ12]. Equation (1) defines the *inter-cluster similarity* between two clusters $C_a \neq C_b$, while the *intra-cluster similarity* of a cluster C is obtained by setting $C_a = C_b = C$.

Figure 2 reports the inter- and intra-cluster similarity scores for each choice of clustering scheme. The absolute similarity numbers are small, due to the sparse nature of tweets and the overall little linguistic overlap within clusters. However, we can see that the intra-cluster similarity is more than twice the inter-cluster similarity, indicating good separation amongst the clusters across all configurations. This seems to confirm that the LDA approach is sufficiently sensitive to discover sub-topics of interest within an already focused general topic, defined by a set of keywords.

The plots in Fig. 3 provide more detailed indication of the contrast between intra- and inter-cluster similarity at the level of detail of individual clusters. For example, in the 4-clusters case, the average of the diagonal values of the raster plot is the intra-cluster similarity reported in Fig. 2, whereas the average of the off-diagonal values represents the inter-cluster similarity. In these plots, darker boxes indicate higher (average) similarity. Thus, a plot where the diagonals are darker than the off-diagonal elements is an indication of a high quality clustering scheme.

Although the similarity metrics are objective and seem to confirm the good quality of the clustering, the plots in Figs. 2 and 3 do not provide much insight into the optimal number of clusters, or indeed their semantic interpretation.

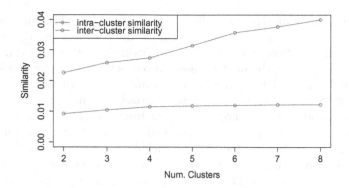

Fig. 2. Intra- and Inter-cluster similarities

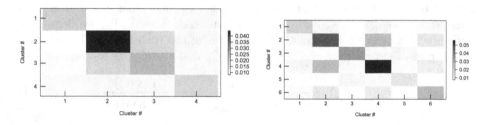

Fig. 3. Inter- and intra-similarity for 4 and 6 clusters topic models

We therefore relied on our domain experts for the empirical selection of the clustering scheme (2,4,6,8 clusters) that would most closely lend itself to an intuitive semantic interpretation of the topics. Their assessment is reported below. In Sect. 4.3 we present a comparison of topics content using our four classes model as a frame of reference.

4.2   Empirical Topics Interpretation

Expert inspection, carried out by native Brazilian Portuguese speakers, considered both the list of words within each topic, and a sample of the tweets for that topic. In this case, the most intelligible clustering scheme had 4 topics. The following is a list of most relevant topics for this scheme:

Topic 1: parado, água, fazer, vacina, até, meu, tão
Topic 2: combate, morte, sáude , confirma, ação , homem, chegar, queda, confirmado, agente
Topic 3: contra, suspeito, sáude , doença, bairro, morrer, combater, cidade, dizer, mutiro
Topic 4: mosquito, epidemia, pegar, foco, casa, hoje, mesmo, estado, igual

---

The importance of the words is given by LDA as a measure of how well they are represented in the topics.[4] Unsurprisingly, topic inspection suggests an

---

[4] Some of the words are just noise. This is due to occasional imperfect lemmatisation during the preprocessing stage.

interpretation that only partially overlaps with the a priori classification we have seen in the supervised case. Specifically, **Topic 1** is closely related to *Jokes*. Most of the tweets for this topic either make an analogy between Dengue and the users lives, or they use the words related to Dengue as a pun. A typical pattern is the following:

> meu [algo como: wpp - WhatsApp, timeline, Facebook, twitter etc] está mais parado do que agua com dengue.
> *My [something like: wpp - WhatsApp, timeline, Facebook, twitter etc] is more still than standing water with dengue mosquito.*

Specific examples include:

> Aitizapi ta com dengue de tão parado (*Aitizapi is so still that it has been infected by dengue*)
> Concessionária tá dando dengue de tão parada que tá (*Car dealership is so still that it has dengue*)

In the first example, the user was playing with the words when referring to the standing status and inactivity in his Whatsapp account. Breeding sites of the Aedes mosquito are mostly found in containers with standing water. In the second, the user is joking about a significant decrease in car purchases due to the emerging economic crisis in Brazil. Many of the jokes in the last epidemic wave have been related to Zika, which in Brazilian Portuguese, has been used as a new slang word for failure or any kind of personal problem.

Topic 2 is interpreted as *news* about increase or decrease of Aedes-borne disease cases as well as specific cases of people who died because of the Aedes-borne diseases, i.e. Dengue, Chikungunya and Zika. It also contains news about the combat of the mosquito in certain locations as well. Examples:

> Rio Preto registra mais de 11 mil casos de dengue e 10 mortes no ano #SP
> *Rio Preto reports more than 11 thousand cases of dengue in the year #SP*
>
> 543 casos estão em análise - Londrina confirma mais de 2,5 mil casos de dengue em 2015 - [URL removed]
> *543 cases of dengue are under analysis - Londrina confirms more than 2.5 cases of dengue in 2015 - [URL removed]*

Topic 3 appears to contain mostly *news about campaigns* or actions to combat or to prevent Aedes-borne diseases, for instance:

> Curcuma contra dengue [URL removed] (*Curcuma against dengue*)
>
> Prefeitura de Carapicuíba realiza nova campanha contra dengue e CHIKUNGUNYA[URL removed]
> *Carapicuíba City Hall launches new campaign against dengue and CHIKUNGUNYA[URL removed]*

The difference between the news in topics 2 and 3 concerns the type of news, which for topic 2 is mostly about the increase or decrease of Aedes-borne diseases, whereas in topic 3 is about campaigns or actions to combat the propagation of the Aedes mosquito.

Finally, **Topic 4** contains mostly *sickness* tweets, with some instances of *jokes*:

> Será que eu to com dengue ? (*I wonder: do I have dengue?*)

4.3   Classes vs Clusters

The point to note in the assessment above is that the most relevant tweets, those corresponding to the *Mosquito Focus* class, are not easily spotted, in particular they do not seem characterise any of the topics. Intuitively, this can be explained in terms of the relative scarcity of these tweets within the stream, combined with the balancing across topics that occurs within LDA.

In order to quantify this intuition, we have analysed the topics content using our pre-defined four classes as a frame of reference. In this analysis, we have used our trained classifier to predict the class labels of all the tweets in the corpus that we used to generate the topics (about 100,000). We then counted the proportion of class labels in each topic, as well as, for each class, the scattering of the class labels across the topics. The results are presented in Tables 2 and 3, respectively, where the dominant entries for each column (resp. row) are emphasised.

It is worth remembering that these results are based on predicted class labels and are therefore inherently subject to the classifier's inaccuracy. Furthermore, the predicted class labels were *not* available to experts when they inspected topic content, thus they effectively performed a new manual classification on a content sample for each topic. Despite the inaccuracies introduced by these elements, Table 2 seems to corroborate the experts' assessment regarding topics 1 and 2, but less so for topics 3 and 4. This may be due to the sampling operated by the experts, which selected content towards the top of the topic (LDA ranks content by relevance within a topic) and may have come across joke entries which are otherwise scarce in topic 4. Although the heavy concentration on joke tweets in topic 1 from Table 2 seems promising (i.e., the other topics are relatively noise-free), Table 3 shows a problem, namely that topic 1 is also where the vast majority of *Mosquito Focus* tweets are found. Thus, although topic 1 segregates the most informative tweets well, it is also very noisy, as these tweets are relatively scarce within the entire corpus.

The analysis just described suggests that topic modelling offers less control over the content of topics, compared to a traditional classifier, especially on

Table 2. Distribution (%) of predicted class labels within each cluster

	Topic 1	Topic 2	Topic 3	Topic 4
News	13.9	**72.6**	27.2	39.4
Joke	**39.5**	0.1	2.8	4.1
Mosquito Focus	30	4.0	12.3	12.5
Sickness	16.6	23.3	**57.7**	**44.0**
Total	100	100	100	100

Table 3. Scattering (%) of predicted class labels across clusters

	Topic 1	Topic 2	Topic 3	Topic 4	Total
News	29.1	28.5	8.9	33.5	100
Joke	**95.0**	0.03	1.05	4.0	100
Mosquito Focus	**79.5**	2.0	5.1	13.4	100
Sickness	34.8	9.1	18.8	**37.3**	100

a naturally noisy media channel. Although relevant content can be ascribed to specific topics, these are polluted by noise. Despite this, LDA performs relatively well on creating sub-topics from a sample that is already focused on a specific topic, such as conversations on the Aedes-transmitted viruses. The main appeal of the classifier is that it makes it straightforward to select relevant content, with acceptable experimental accuracy. In our follow on research we are investigating ways to combine the benefits of the two approaches. Specifically, we are studying a unified semi-supervised model where topic modelling can be used to improve the accuracy of the classifier, i.e., by automatically expanding the training set, and to alleviate the cost of re-training at the same time.

5   Summary

In this paper we discussed methods for detecting relevant content in a Twitter stream that has been pre-filtered to focus on a specific topic, in this instance online discussions around Dengue and other Aedes-borne diseases in Brazil. Relevance is defined operationally in terms of four classes within the broad topic of Dengue. When reliably segregated from noise, relevant content can be used in multiple ways in the context of health vigilance to combat epidemics caused by the Aedes mosquito. We have compared two approaches for detecting relevance, supervised classification and clustering by topic modelling. Our experimental results indicate that the clusters produced using topic modelling tend to be noisy, perhaps because LDA is not very effective on text content that is pre-filtered for a specific set of keywords. Supervised classification, on the other hand, is costly as manual annotation requires multiple rounds due to ambiguities in the content, but is more appealing as a good proportion of actionable messages are segregated, i.e., in the two most relevant classes. We are currently exploring ways to combine the two approaches into one semi-supervised model, i.e., by exploiting the topics to enhance the training set and alleviate the cost of re-training.

Acknowledgments. This work has been supported by MRC UK and FAPERJ Brazil within the Newton Fund Project entitled *A Software Infrastructure for Promoting Efficient Entomological Monitoring of Dengue Fever*. The authors would like to thank Oswaldo G. Cruz (Fundao Oswaldo Cruz, Programa de Computacao Cientifica) and Leonardo Frajhof (Unirio, Rio de Janeiro, Brazil) for their contributions to this paper, and Prof. Wagner Meira Jr. and his team for sharing their 2009–2011 Twitter datasets [GVM+11].

References

[ALG+11] Achrekar, H., Lazarus, R., Gandhe, A., Yu, S., Liu, B.: Predicting flu trends using Twitter data. In: IEEE Conference on Computer Communications Workshops (INFOCOM WKSHPS), pp. 702–707. IEEE (2011)

[AZ12]   Aggarwal, C.C., Zhai, C.X.: A survey of text clustering algorithms. In: Aggarwal, C.A., Zhai, C.X. (eds.) Mining Text Data, pp. 77–128. Springer, New York (2012)

[BNG11] Becker, H., Naaman, M., Gravano, L., Topics, B.T.: Real-world event identification on Twitter. In: Proceedings of ICWSM, pp. 1–17 (2011)

[BNJ03] Blei, D.M., Ng, A.Y., Jordan, M.I.: Latent Dirichlet allocation. J. Mach. Learn. Res. **3**, 993–1022 (2003)

[CDC15] CDC. Centers for Disease Control and Prevention (2015). http://www.cdc.gov/dengue/. Accessed 15 Dec 2015

[CW14]   Cheng, T., Wicks, T.: Event detection using Twitter: a spatio-temporal approach. PloS One **9**(6), e97807 (2014)

[DSL+11] Dela Rosa, K., Shah, R., Lin, B., Gershman, A., Frederking, R.: Topical clustering of tweets. In: SIGIR 3rd Workshop on Social Web Search and Mining (2011)

[GBH09] Go, A., Bhayani, R., Huang, L.: Twitter sentiment classification using distant supervision. CS224N Project Rep. Stanford **1**(12), 12 (2009)

[GVM+11] Gomide, J., Veloso, A., Meira, W., Almeida, V., Benevenuto, F., Ferraz, F., Teixeira, M.: Dengue surveillance based on a computational model of spatio-temporal locality of Twitter. In: Proceedings of the ACM WebSci 2011, Koblenz, Germany, 14–17 June 2011, pp. 1–8 (2011)

[LC10]   Lampos, V., Cristianini, N.: Tracking the flu pandemic by monitoring the social web. In: 2nd International Workshop on Cognitive Information Processing, CIP 2010, pp. 411–416 (2010)

[MPLC13] Morstatter, F., Pfeffer, J., Liu, H., Carley, K.: Is the sample good enough? Comparing data from Twitter's streaming API with Twitter's firehose. In: Proceedings of ICWSM, pp. 400–408 (2013)

[NGS+09] Nagarajan, M., Gomadam, K., Sheth, A.P., Ranabahu, A., Mutharaju, R., Jadhav, A.: Spatio-temporal-thematic analysis of citizen sensor data: challenges and experiences. In: Vossen, G., Long, D.D.E., Yu, J.X. (eds.) WISE 2009. LNCS, vol. 5802, pp. 539–553. Springer, Heidelberg (2009)

[PR15]   PUC-Rio. Efficient Monitoring of Aedes Mosquito in Brazil (2015). vazadengue.inf.puc-rio.br/. Accessed 15 Dec 2015

[RDL10] Ramage, D., Dumais, S.T., Liebling, D.J.: Characterizing microblogs with topic models. ICWSM **10**, 1 (2010)

[REC12] Ritter, A., Etzioni, O., Clark, S.: Open domain event extraction from Twitter. In: Proceedings of the 18th ACM SIGKDD International Conference on Knowledge Discovery and Data Mining - KDD 2012, p. 1104 (2012)

[Rou87] Rousseeuw, P.J.: Silhouettes: a graphical aid to the interpretation and validation of cluster analysis. J. Comp. Appl. Math. **20**, 53–65 (1987)

[SOM10] Sakaki, T., Okazaki, M., Matsuo, Y.: Earthquake shakes Twitter users: real-time event detection by social sensors. In: Proceedings of WWW 2010, p. 851 (2010)

[WLJH10] Weng, J., Lim, E., Jiang, J., He, Q.: Twitterrank: finding topic-sensitive influential Twitterers. In: Proceedings of WSDM 2010, pp. 261–270. ACM (2010)

Experimental Measures of News Personalization in Google News

Vittoria Cozza[1,2]([✉]), Van Tien Hoang[3], Marinella Petrocchi[1],
and Angelo Spognardi[1,4]

[1] IIT-CNR, Pisa, Italy
{v.cozza,m.petrocchi,a.spognardi}@iit.cnr.it
[2] DEI, Polytechnic University of Bari, Bari, Italy
[3] IMT School for Advanced Studies, Lucca, Italy
vantien.hoang@imtlucca.it
[4] DTU Compute, Lyngby, Denmark

Abstract. Search engines and social media keep trace of profile- and behavioral-based distinct signals of their users, to provide them personalized and recommended content. Here, we focus on the level of web search personalization, to estimate the risk of trapping the user into so called Filter Bubbles. Our experimentation has been carried out on news, specifically investigating the Google News platform. Our results are in line with existing literature and call for further analyses on which kind of users are the target of specific recommendations by Google.

Keywords: Filter bubbles · Web search results · News publishers

1 Introduction

Search engines and social media provide Internet users the opportunity to discuss, get informed, express themselves and interact for a myriads of goals, such as planning events and engaging in commercial transactions. The management of relational networks, e.g., for researches, data gathering and sharing, raises significant questions about the quality of the information retrieved. In his popular work on Filter Bubbles [13], Pariser was one among the first ones to theorize the phenomenon according to which users are unknowingly trapped in "protective" bubbles, created by search engines and social platforms to automatically filter contents. As an example, the author reports how some posts gradually disappeared from his Facebook news feed, probably driven by his historical navigation activities once logged into the popular social platform.

Remarkably, while the most active users of social media act as gatekeepers, the new guardians of information, by personalizing its spread, the same media and search engines track the user navigation and filter the search results. In such

Partly funded by the Registro.it project *MIB* (My Information Bubble).

S. Casteleyn et al. (Eds.): ICWE 2016 Workshops, LNCS 9881, pp. 93–104, 2016.
DOI: 10.1007/978-3-319-46963-8_8

a way, gatekeepers and social media become "dangerous intermediaries" [12], with the natural potential consequence of narrowing the world view. Confined in comfortable micro-arenas, the potential risk is loosing the communicative potential of the web, in which information management is performed in a bottom-up fashion [8]. The practice of filtering re-ordering is testified by the service providers themselves. As an example, the Google patent on personalization of web search [10] states the existence of mechanisms linking the re-order of search results to the preferences in the user profiles. In last recent years, Academia spent a significant effort in measuring the level of such personalization, giving raise to seminal work like the one in [6] on Google, which showed that criteria mostly influencing personalization of search results are geo-localization and login into the platform.

Personalization affects also online advertising. Indeed, recently, the traditional advertising approach has moved towards a targeted one: the ad is shown only to online users with a specific profile - location, gender, age, e-shopping history are among the monitored aspects. Although personalised ads have the significant advantage to guide the customer mostly towards products she likes, concerns were born since the ads system could (1) hide to the user other potential interesting products [13]; and (2) expose user private information [1].

In this paper, we focus on news aggregators. The target of our analysis is Google News and the goal is to measure the level of personalization of results returned to different kinds of users when they search over a news dataset. Google News offers a panoramic view on several articles on different subjects, redirecting the users on publishers' accounts, when they can select a news to read. The proposed articles range over a variety of topics like business, technology, entertainment, sports and many others. Moreover, the platform offers the capability to personalize the kind of news shown both to logged and not logged users, based, e.g., on the frequency of news sources, and on specific topics.

We consider two aspects of news personalization, defined by the related literature as *expected* and *unexpected passive personalization* [3]. Expected (resp., unexpected) personalization is an explicit (resp., not claimed) personalization, described (resp., undocumented) by Google in reports and other documentations. The term *passive* means that such personalization has not been directly configured by the user through the appropriate functionalities offered by Google News.

To measure personalization, we compare the results provided by Google News to logged users, which we previously trained with different searches and visits to websites, for a set of topics and publishers.

For the user training, we adopt as the reference dataset the *Signal Media One-Million News Articles* dataset, which is a public collection of articles, to serve the scientific community for research on news retrieval. In particular, we restrict and focus only on those elements in the dataset with source espn.com (Entertainment & Sports Programming Network), with more than 7,000 news present in the dataset. We only pick this publisher since it is linked to Google News and it has a large number of news in English. Moreover, espn.com is also part of the Google Display Network (GDN), namely a large set of websites publishing Google advertisements[1], that is publicly known to make use of user profiling to provide targeted advertisements [7].

To train the users, we extract the *relevant entities*, such as organizations, persons, and locations, mentioned in the titles of the espn.com news in the dataset and we use these entities to emulate the behaviors of users interested in *Sports*. We, then, exploit such behaviors to investigate either expected and unexpected passive personalization over Google News.

The intuition behind our methodology is that intensively engaging a user over a specific publisher and topic would lead Google News to infer a specific interest of such user for that publisher and topic. Thus, successive search queries of that user could lead the provider to alter the order of the results, e.g., ranking first the news of the specific publisher, with respect to the order of the news provided to a user without past activity.

While we observe expected personalization in the dedicated Suggested for You (SGY) section over Google News, there are not sensitive differences in the news results shown on the main page, between the trained user and a fresh one. This leads to results in line with related work in the area, such as [3], achieved however through a different experimental setting. The current study contributes to (1) evaluate if news are sorted with or without regard to past behaviors of the user, and (2) define the settings within which users are not exposed to a news results order exchange. The last aspect is particularly relevant since studies in the literature have shown how users are deeply influenced by the results shown by search engines in return to their queries, see, e.g., [15].

The rest of the paper is structured as follows. Next section briefly relates on Google News news personalization. In Sect. 3, we introduce the reference dataset we start from in our experiments. Section 4 describes the techniques used to extract relevant entities from the *Signal Media* news titles. Section 5 presents settings and implementation of the training and the test phase over Google News and gives experimental results. Finally, Sect. 6 concludes the paper.

2   Google News Personalization

Google has provided many details about the mechanisms it uses for personalization. For example, a first kind of personalization exists for logged users with their web history activated. On the specific personalization over Google News, work in [11] describes an enhanced recommender system to offer in the Suggested for You (SGY) section of Google News news, customized such as to be closely inherent to the users' interests.

However, there exist other forms of personalization, based also on "past news browsing information", as explained in the Google support documentation on news personalization, available at https://support.google.com/news/answer/3010317?hl=en. The effects of such personalization could be subtle, since most

---

[1] https://support.google.com/adwords/answer/2404190?hl=en All URLs have been accessed on May, 15, 2016.

users are not aware of its existence, and, thus, quite obviously, they do not know how to disable it, if needed.

The personalization based on past news browsing has been subject of the study in [3], where the authors define two properties: expected and unexpected passive personalization, where passive in both cases means that user has not customized a personalized search through the functionalities of Google News. To measure both kinds of personalization, the user searched over Google News specific topics and publishers. Then, the same user connected to Google News again, and both SGY and the home page are analyzed, trying to find evidence of the user previous activity during the training phase. Results of [3] showed effects for the expected passive personalization (the one supposedly affecting the news in the SGY section) and no effect for the unexpected one (possibly affecting the news shown in the main section of the Google News home page).

In the following, we will show design and implementation of a set of experiments to evaluate passive personalization, with however different settings with respect to what done in [3]. Indeed, we analyze results of queries made by the users, rather than analyzing the news shown by default by Google News once connected to the platform.

In [3], the authors also train both logged and not logged users on a set of specific publishers (USA Today, Reuters, the Wall Street Journal, the Economist). A novelty of our approach is that, by letting users search keywords from real news published by espn.com as they appear on the *Signal Media* dataset, we are pretty sure to find indeed news about that publisher as the results of the searches over Google News, not only during the training phase, but also during the test phase. Instead, by focusing on news in the main page without searching for some particular news, work in [3] suffered from the limitation that the chosen publisher could not have been present, given the intrinsic volatility of news.

3   Reference Dataset

For our experiments, we refer to the *Signal Media One-Million News Articles* dataset[2]. It consists of a variety of news (from different sources) collected over a period of one month, from September 1st, 2015. Overall, the dataset counts 1 million articles, mainly in English. The sources for these articles include major web sites, such as espn considered in this paper, as well as local news sources and blogs. Each article consists of its unique identifier, the title, the textual content, the name of the article source, the publication date, and the kind of the article (either a news or a blog post).

The dataset counts 93k individual unique sources, 265,512 Blog articles, and 734,488 news articles. Moreover, at the time of our study the first five most recurrent sources in the dataset are MyInforms (19,228 occurrences), espn (from the main website and affiliated ones: (7,713), Individual.com (5,983), 4Traders

---

[2] http://research.signalmedia.co/newsir16/signal-dataset.html.

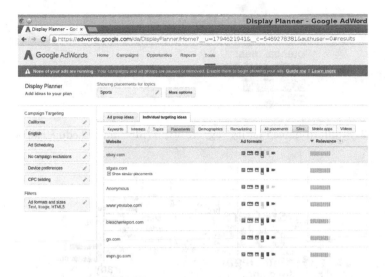

Fig. 1. Google display planner: `espn` belongs to the google display network.

(4,438), NewsR.in (4,039), and `Reuters` (3,898). We have chosen `espn` since it is a very large source in the dataset, it has the articles linked to Google News and, finally, because we know in advance that the user activity on this website is tracked by Google. Indeed, `espn` belongs to the Google Display Network (Fig. 1).

4   Named Entity Extraction

Named Entity Recognition (NER) is the process of identifying and classifying entities within a text. In the news domain under investigation, common entities are persons, locations and organizations. NER state-of-the-art systems use statistical models (i.e., machine learning) and typically require a large amount of manually annotated training data in combination with a classifier. The best solutions, in terms of classification accuracy, are generally based on conditional random fields (CRF) classifiers [5]. For our experiment, we use the Stanford NER tagger to extract the entities from the news titles in the *Signal Media* dataset. The tagger is indeed implemented through a linear chain CRF sequence labeler classifier and it is part of the Stanford Core NLP[3]. Thus, we exploit the Natural Language Toolkit NLTK[4], which processes natural language through Python programming. NLTK gives the access to the Stanford tagger and to valuable linguistic resources and data models. In detail, as learning model for the classifier, we adopt a ready-to-use model by Stanford, called english.all.3class.distsim.crf.ser.gz, available in NLTK. Figure 2 shows an example of entities extracted from the titles of `espn` news, while Fig. 3 shows the most recurrent entities under the form of a word cloud.

---

[3] http://stanfordnlp.github.io/CoreNLP/.

[4] http://www.nltk.org/.

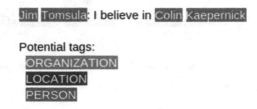

Fig. 2. Stanford named entity tagger: http://nlp.stanford.edu:8080/ner/

Fig. 3. The most recurrent entities in titles with source `espn`.

On a total of more than 7,000 news titles from `espn` and affiliated sub-publishers, we have considered the ones from the main website (espn.com), consisting of 577 titles. We have found 662 entities and 465 unique ones.

5   Experiments

The final goal of the experiments is measuring how users past online behavior affects the news provided by Google News. We investigate two properties, expected (EPP) and unexpected (UPP) passive personalization, as in [3]. The expected version of passive personalization has been disclosed by Google in [11]. Citing Google, "If a user signs in to her Google Account and explicitly enables Web History, the system will record her click history and generate a personalized section for her, named Suggested for [account], containing stories recommended based on her click history in Google News". Instead, UPP is supposed to have an effect on news shown on the main section of the Google News home page.

Several frameworks have been proposed for analyzing personalized search results and advertisements on Google. The interested reader can refer to [4] for a full survey of measurement tools and their comparison. We have chosen to adopt AdFisher[5] [14], a tool to analyze interactions between online user behaviors, advertisements shown to the user, and advertisements settings. Later, AdFisher

---

[5] https://github.com/tadatitam/info-flow-experiments.

has been extended for handling Google searches and news searches and measuring personalization in query results, see, e.g., [3]. Further, a branch version has been created for handling product searches in Google Shopping, to measure price steering [9]. AdFisher has also been used for statistical evaluations, e.g., to measure how users are exposed to Wikipedia results, in return to their Google web searches, see, e.g., [2].

In our work, AdFisher runs browser-based experiments that emulate search queries and basic interactions with the search results, i.e., click only search results satisfying a certain property, e.g., coming from a specific publisher. In particular, AdFisher is automatized with Selenium[6] to efficiently create and manage different users with isolated Firefox client instances, each of them with their own associated cookies, to enable the personalization from the server.

For measuring both EPP and UPP, we emulate two real users, logged into Google, connecting to two separate browser instances, one for each user. Since we are investigating a publisher that is commonly and mostly accessed from US[7], all our experiments are with the users connected from US. We have used the Digital Ocean VPS Service Digital[8] to gain access to machines located in the US.

5.1 Unexpected Passive Personalization

The following list describes the experiments for training one of the two users and for comparing, in the test phase, the obtained search results with those of the other user.

The user is trained as follows:

- She visits sports-related GDN website pages, including pages from espn.com. The websites have been selected with Google Display Planner—a tool providing a series of websites taking part to the GDN and linked to specific topics.
- She issues several queries on Google News, using as keywords the entities extracted from the *Signal Media* dataset (see Sect. 3).
- She clicks only on the results of the news with source espn.com and spends some time on the linked page.

The training phase lasted about eight hours. To evaluate UPP for both the users, we have performed searches on Google News, with different keywords with respect to the training phase. Both the users searched for 32 test keywords. We recall that the second user does not undergo any training phase.

As highlighted by the Google News guide[9], a form of personalization exists even for users not logged into a Google account. For these users, the "Google News experience will be personalized based on past news browsing information". We are indeed interested in that kind of personalization, based on the previous online behavior.

---

[6] http://www.seleniumhq.org/.
[7] http://www.alexa.com/siteinfo/espn.com.
[8] https://www.digitalocean.com.
[9] https://support.google.com/news/answer/3010317?hl=en.

Table 1 reports the training details.

Table 1. Training behavior

Visited pages	Searched keywords	Read articles	Avg time on website	Location
17	464	100	50 s	New York, USA

Noticeably, for all the test queries, the fresh user and the trained user have been shown exactly the same results in the main section of Google News home page. Thus, we were unable, under our experiment context, to reveal personalization based on past news browsing information, contrary to what claimed by Google, in the Google News guide mentioned above.

5.2   Expected Passive Personalization

To analyze the expected passive personalization, we focus on the Google News SGY section. Given a fresh logged-in user, Google News does not provide the user with such a section. Indeed, the user needs to have formerly interacted with Google (either Google search or Google News engine). We follow two approaches to make that section appear, as described in the following.

1. In the first approach, we try to build two user profiles interested in traveling, letting both the users visit 30 travel-related websites, searching for 327 travel-related keywords on Google News, and clicking on the first result. We have chosen the topic "travel" since we consider it a topic quite disjoint from sport. We have emulated such a behavior until the SGY section appeared. This happened after four iterations of searching keywords and visiting websites. This pre-training phase lasted four days. Remarkably, when the SGY section was populated, it was just with one entry, related to a crime news about a murder in Las Vegas, having nothing in common with traveling. The travel-related keywords and websites were obtained with the support of the Google Display Planner.
2. In the second approach, we try to build two user profiles interested in sports. For both the users, we have trained them according to the training behavior described in Sect. 5.1. In this case, the SGY section appears earlier, probably because of a larger interaction of the user with the browser. Indeed, the user clicks on all the news from the espn publisher, while, for the travel scenario, she was clicking only on the first result, per search. After one day, using the training settings in Table 1, we obtain a SGY section with ten news. A visual examination helps us to assess that such news are related to Sport. Although a text mining approach would be have more effective in assessing topics of such news, we can argue that the read and click behavior thought for users interested in sports has been appropriately designed.

From the results of this preliminary step, we envisage that the number of SGY news depends on the interactions of the user with Google. Indeed, different interactions yield to different results, both for number and content of news in that section.

The travel topic was not a winning choice. In fact, we observed the lack of timely news related to this topic. This let us argue that the implemented behavior does not lead to a profile of a user really interested in travels. Thus, hereafter, we concentrate the experiments on users trained on sports.

In the following, we consider only two logged users, interested in sports, each with a SGY section. We will further investigate if different behaviors of such users yield to distinct results in that section.

- Training: for the first user, we have repeated the training described in Sect. 5.1. The second user just waits.
- Test: for both users, we have refreshed the Google News home page every 10 min to capture real-time events (default reload time of Google News is 15 min[10]) and we have focused on the SGY section only.

Both the pre-training session (to let the SGY section appear) and the training phase lasted eight hours. Thus, we have tried to answer the following: *do two users, both interested in sports, but with different interactions with sport websites and sports news, have different SGY news?*

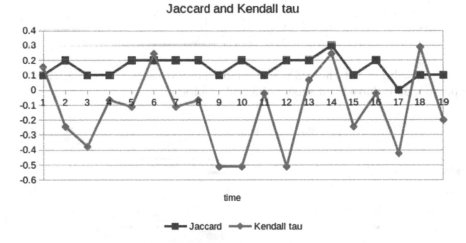

Fig. 4. Jaccard and Kendall indexes of the SGY sections.

We have computed standard metrics usually adopted to measure web search personalization (Jaccard and Kendall indexes). Given two sets P and Q, Jaccard Index is 1 when the sets are identical and 0 when their intersection is empty while Kendall index (Kendall tau) quantifies the correlation between P and Q.

---

[10] https://news.google.com/news/settings.

This index ranges from -1 to 1, where 0 means no correlation, 1 means same order and -1 reverse order.

In Fig. 4, we can see the differences in the SGY sections of two users, according to the introduced metrics. The sections have from 1 to 3 identical news, out of ten (Jaccard index is from 0 to 0.3). Most of the time, the order of these news is not correlated (for 13 out of 19 the Kendall index is negative). It is worth noting that the untrained user obtains more real-time news (updated within 60 min) than the other one. Even if we have no clear evidence to explain this phenomenon, we can suppose that untrained users have a wider interest domain than the trained one, leading to a wider suggestion of news by Google.

As a corollary, we have also considered how many news, among the suggested ones, are from espn.com. The results are in Table 2, at each refreshing time of the SGY section. Overall, the trained account has been shown more espn news than the untrained one. Considering the top 10 news shown to each user, we always got 1 to 3 espn news in their SGY section. It is worth however noting that the result is also related to the number of espn news actually published at experimenting time. Indeed, SGY section tends to show first the most recent news.

Table 2. Statistic of news from espn.com in SGY section, at each refreshing time

Time slot	t1	t2	t3	t4	t5	t6	t7	t8	t9	t10	t11	t12	t13	t14	t15	t16	t17	t18	t19	All
Untrained account	1	1	1	0	1	1	1	1	0	1	0	1	1	3	1	1	1	1	2	**19**
Trained account	2	1	1	1	2	1	2	1	1	2	1	1	1	2	1	1	1	1	2	**25**

Figure 5 highlighs the SGY section of the two users, at testing time. As expected, users have been shown different results.

Fig. 5. Snapshot of the two SGY sections: differences and similarities. Snapshot of the two SGY sections: differences and similarities.

6 Conclusions

In this paper, we focused on news personalization on Google News, aiming at measuring the level of personalization (claimed by Google itself), under different contexts: logged users, expected (in SGY sections) and unexpected (in Google News home) personalization. We differ from related work in the literature mainly because we observe the results obtained in return to specific user queries. However, at least for our experiments configuration, we did not observe particular differences in the results obtained by a trained user and a fresh one. Instead, we found interesting results when we searched for expected personalization, looking specifically at the SGY section of Google News. Since the section is not automatically shown to non logged users, nor to freshly logged ones, we carried on experiments to let the section appear on the users pages. Our approach showed that, depending on the kind and number of interactions a user has on the platform, the SGY section differs both in content and number of the shown news. Furthermore, results after training a specific user over a particular topic leads to a different SGY section with respect to SGY of the non trained user (confirming, in this case, previous related results).

References

1. Conti, M., Cozza, V., Petrocchi, M., Spognardi, A.: TRAP: using targeted ads to unveil google personal profiles. In: IEEE International Workshop on Information Forensics and Security, pp. 1–6 (2015)
2. Cozza, V., Hoang, V.T., Petrocchi, M.: Google web searches and wikipedia results: a measurement study. In: Proceedings of 7th Italian Workshop on Information Retrieval (IIR) 2016, Venice, Italy (2016). CEUR Workshop Proceedings http://ceur-ws.org
3. Datta, A., Datta, A., Jana, S., Tschantz, M.C.: Poster: information flow experiments to study news personalization. In: 2015 IEEE 28th Computer Security Foundations Symposium (CSF). IEEE (2015)
4. Englehardt, S., Narayanan, A.: Online tracking: a 1-million-site measurement and analysis. Technical report, May 2016
5. Finkel, J.R., Grenager, T., Manning, C.: Incorporating non-local information into information extraction systems by Gibbs sampling. In: Proceedings of the 43rd Annual Meeting on Association for Computational Linguistics, ACL 2005, pp. 363–370. Association for Computational Linguistics, Stroudsburg (2005)
6. Hannak, A., Sapiezynski, P., Kakhki, A.M., Krishnamurthy, B., Lazer, D., Mislove, A., Wilson, C.: Measuring personalization of web search. In: 22nd International Conference on World Wide Web, WWW 2013, pp. 527–538 (2013)
7. Haveliwala, T., Jeh, G., Kamvar, S. Targeted advertisements based on user profiles and page profile. US Patent 8,321,278, 27 November 2012
8. Hindman, M.: The Mith of Digital Democracy. Princeton University Press, Princeton (2009)
9. Hoang, V.T., Cozza, V., Petrocchi, M., De Nicola, R.: Online user behavioural modeling with applications to price steering. In: Proceedings of the 2nd International Workshop on Personalization and Recommender Systems in Financial Services (FINREC 2016), Bari, Italy, pp. 16-21 (2016). CEUR Workshop Proceedings http://ceur-ws.org/Vol-1606/paper04.pdf. ISBN: 1613-0073

10. Lawrence, S.: Personalization of web search, US Patent App. 10/676,711, March 2005
11. Liu, J., Dolan, P., Pedersen, E.R.: Personalized news recommendation based on click behavior. In: Proceedings of the 15th International Conference on Intelligent User Interfaces, IUI 2010, pp. 31–40. ACM, New York (2010)
12. Morozov, E.: The Net Delusion: The Dark Side of Internet Freedom. Perseus Books, Cambridge (2011)
13. Pariser, E.: The Filter Bubble: What the Internet Is Hiding From You. Penguin Group, London (2011)
14. Tschantz, M., Datta, A., Datta, A., Wing, J.: A methodology for information flow experiments. In: 2015 IEEE 28th Computer Security Foundations Symposium (CSF), pp. 554–568, July 2015
15. Zanker, M., Ricci, F., Jannach, D., Terveen, L.: Measuring the impact of personalization, recommendation on user behaviour. Int. J. Hum.-Comput. Stud. **68**(8), 469–471 (2010)

1st International Workshop on Liquid Multi-Device Software for the Web (LiquidWS 2016)

XD-Bike: A Cross-Device Repository
of Mountain Biking Routes

Maria Husmann$^{(\boxtimes)}$, Linda Di Geronimo, and Moira C. Norrie

Department of Computer Science, ETH Zurich, 8092 Zurich, Switzerland
{husmann,lindad,norrie}@inf.ethz.ch

Abstract. Despite the high level of interest in cross-device interaction, only few fully functional example applications exist. With XD-Bike, we built one of our own to not only showcase the use of our cross-device framework but also gather insights into the development process. XD-Bike is an application for mountain bikers that is built with web technologies and adapts to the set of devices at hand. The user interface is distributed across all available devices taking into account the space requirements of the UI elements, their importance, and the available space on devices. While using the cross-device framework eased the development process, we found the testing and debugging challenging due to the distributed nature of the application and the large set of possible device combinations.

Keywords: Distributed user interface · Cross-device · Liquid applications · Web

1 Introduction

Cross-device or liquid applications [1] that adapt to the set of devices at hand have received a lot of attention from the research community in recent years. The user interfaces of such applications spread across multiple devices and interaction with one device affects paired devices. Frameworks [2,3] and design tools [4,5] have been built to facilitate the development, but few example applications exist. Our goal was to build a complete example application as a showcase for our own web-based framework XD-MVC[1] which is based on the MVC pattern [6]. Furthermore, we wanted to explore the development process of cross-device applications to find difficulties and gather requirements for better tools. As an application domain, we chose a tour repository for mountain bikers and named the project XD-Bike.

Planning a mountain bike tour can be challenging. For the tour to be successful and fun, many factors come into play. The technical difficulty should not be too high (frustrating) or too low (boring). It should neither be too short nor too long. The same goes for the elevation gain. Not only is the total elevation gain

---

[1] https://github.com/mhusm/XD-MVC.

© Springer International Publishing AG 2016
S. Casteleyn et al. (Eds.): ICWE 2016 Workshops, LNCS 9881, pp. 107–113, 2016.
DOI: 10.1007/978-3-319-46963-8_9

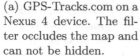

(a) GPS-Tracks.com on a Nexus 4 device. The filter occludes the map and can not be hidden.

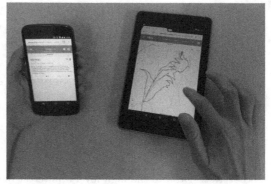

(b) XD-Bike on a Nexus 4 and a Nexus 7 device showing a map and corresponding summary. The map is shown on the larger device.

Fig. 1. An existing mountain bike repository and XD-Bike.

relevant, but also the grade of the ascent - too steep and it is unridable. Last but not least, the location of the tour is an important factor. Even though most riders will be faced with these questions and challenges during planning, there is no one-size-fits all solution. Technical skills and endurance vary among bikers. Also, the form on the day of the riders and weather conditions may require plans to be changed at short notice.

These challenges have been recognised and there are books and magazines that propose routes, special bike maps, and online route repositories that can help bikers plan their tours. In terms of online repositories for Switzerland, two popular examples are Mountainbikeland[2] and GPS-Tracks.com[3]. Both offer a fully featured desktop website and an app for mobile devices. The apps offer only a reduced set of features compared to the website. GPS-Tracks.com has recently been changed to a responsive design, but its use on small screens is still problematic (Fig. 1(a)). It certainly is challenging to present such a wealth of information with limited screen real estate.

Yet, a common scenario is that a couple of riders spend several days in a hotel in the mountains with limited equipment. To keep luggage light, notebooks are often left at home and using just the apps may be inconvenient. However, typically each rider will have at least a smartphone. In addition, one or multiple tablets might be present and a modern hotel may offer a smart TV in the rooms. This results in a scenario where a cross-device application could be beneficial to the users. There is the need to display a lot of information and individually none of the devices are well suited. The phones have small screens and the TV, if present at all, typically has limited input capabilities. But taken together, these

---

[2] http://www.mountainbikeland.ch.

[3] http://gps-tracks.com/.

devices offer increased screen real estate and input capabilities that XD-Bike takes advantage of by distributing its user interface across all of them (Fig. 1(b)).

2 Design

XD-Bike contains a set of mountain bike routes that can be accessed from either a list view or a map. For both views, a filter is available that allows the tracks to be selected according to a set of criteria including difficulty, location, beauty, and keywords (Fig. 2(a)). There is a detailed view for each track where additional information such as a textual description, a map, elevation profile, and pictures are displayed (Fig. 2(b)).

(a) The list view with the filter.   (b) The detailed view on a single device.

Fig. 2. XD-Bike design

Multiple devices can be paired and used in combination thereafter. The pairing is done by either sharing a URL, for example using Android Beam[4] or an instant messenger, or by manually entering the ID of a device. Figure 3 shows the pairing view where all connected devices are listed. A number in the top toolbar that is constantly displayed shows how many devices are currently paired. This provides feedback to the user and lets them verify if the pairing process was successful or could indicate if a device is disconnected. Once the devices are paired,

---

[4] https://support.google.com/nexus/answer/2781895?hl=en.

Fig. 3. The pairing menu, showing the pairing URL (1), the number of paired devices including this one (2), and a list of all paired devices (3).

the first device is assigned the role of a *controller* and all other devices are *viewers*. Only the controller can select a route for which details will be displayed on all paired devices.

The detailed view is made up of multiple tiles or *fragments*. Each fragment contains a coherent unit of information for a chosen track, for example a map or an elevation profile. When multiple devices are paired, these fragments are distributed across the devices. The distribution is determined by an algorithm that takes into account the available space on the device, the space requirements of the fragment and the priority of the fragment. For example, the map fragment needs the most space and has a high priority. The track summary has a low space requirement but the highest priority, while the additional info fragment has both a low space requirement and low priority. Devices can display one or multiple fragments depending on their size. In a first step, we calculate how many fragments can be displayed over all devices. We then sort the fragments by priority and discard those with the lowest priority that would not fit on any device. In the next step, the remaining fragments are distributed across

Fig. 4. XD-Bike on three small devices. Each device displays one fragment. The leftmost device is the controller which is indicated by an icon at the top right next to the number of connected devices.

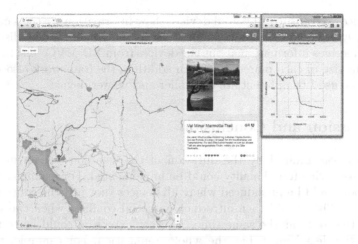

Fig. 5. XD-Bike on a small and a large device. The large device shows a large fragment (map) and two smaller ones.

all devices giving larger fragments the bigger slots. This computation is done client-side on each client. To ensure that the algorithm is deterministic and that each device reaches the same global distribution, unique device IDs assigned by XD-MVC are used to achieve a total order.

Figure 1(b) shows the application distributed across two small devices. Each device displays one fragment. The summary and the maps have the highest priority and are thus displayed. The map requires more space than the summary. Consequently, the map is displayed on the larger of the two devices. Similarly, Fig. 4 shows the interface distributed across three small devices and Fig. 5 across a large and a small device. The large device displays three fragments with the maps assigned to the largest space. Each time a new device is paired, the system re-evaluates the available space and re-distributes the fragments. On each device, the user can manually override the distribution and choose a fragment to be displayed on the device.

On the controller device, routes can be selected from the list view, but navigating back and forth between the list and the detail view could be tedious. To make route selection easier, we also implemented tilt-and-tap interactions [7]. To go the next or previous route on the list, the user can quickly tilt the controller device to the right or to the left.

3   Implementation

XD-Bike was implemented as a semester project by three master students using our own cross-device framework XD-MVC and the Polymer[5] library. XD-MVC handles the device to device communication and data synchronisation among the

---

[5] https://www.polymer-project.org/.

paired devices. To ensure all paired devices show information for the same route, the currently selected route is configured to be synchronised. The distribution algorithm receives information about the devices from XD-MVC. The framework provides the size of each paired device. In addition, the framework also supports roles that we used to differentiate controller and viewer devices.

4   Discussion and Conclusion

The project was limited by the duration of the semester. Given more time, further cross-device functionality could be implemented. For example, a collaborative mode could be introduced where all devices (not just the controller) can select routes. This would then introduce potential conflicts when two users try to change the route at the same time. Another extension could be to not only distribute the detail view, but the whole application. For example, the filter could be moved to a small device and the map showing the results to a larger one. How the users can be informed of the possibilities available in a cross-device application and how they can use all their devices also remains a question that could be addressed in the future.

The project showed that having a cross-device framework facilitates the development. Most of the time was spent on the general implementation and only a small part was spent on integrating with the cross-device framework. At the same time, our experience showed that developing a cross-device application is still challenging. As the application is web-based, browser tools were used to debug the application and multiple browser windows simulated multiple devices. Occasionally, we also tested the application on real mobiles and tablets. This process is rather tedious and devices have to be paired again every time the application is reloaded. This has sparked us to explore better tools for testing and debugging cross-device applications [8].

Acknowledgements. This project was supported by grant No. 150189 of the Swiss National Science Foundation (SNF). We would like to thank Dhivyabharathi Ramasamy, Alexander Richter, and Marko Zivkovic for their contributions to this project.

References

1. Mikkonen, T., Systä, K., Pautasso, C.: Towards liquid web applications. In: Cimiano, P., Frasincar, F., Houben, G.-J., Schwabe, D. (eds.) ICWE 2015. LNCS, vol. 9114, pp. 134–143. Springer, Heidelberg (2015)
2. Yang, J., Wigdor, D.: Panelrama: enabling easy specification of cross-device web applications. In: Proceedings of the CHI (2014)
3. Badam, S.K., Elmqvist, N.: PolyChrome: a cross-device framework for collaborative web visualization. In: Proceedings of the ITS (2014)
4. Nebeling, M., Mintsi, T., Husmann, M., Norrie, M.C.: Interactive development of cross-device user interfaces. In: Proceedings of the CHI (2014)

5. Husmann, M., Nebeling, M., Pongelli, S., Norrie, M.C.: MultiMasher: providing architectural support and visual tools for multi-device mashups. In: Benatallah, B., Bestavros, A., Manolopoulos, Y., Vakali, A., Zhang, Y. (eds.) WISE 2014, Part II. LNCS, vol. 8787, pp. 199–214. Springer, Heidelberg (2014)
6. Krasner, G.E., Pope, S.T.: A cookbook for using the model-view controller user interface paradigm in Smalltalk-80. J. Object Oriented Program 1(3), 26–49 (1988)
7. Di Geronimo, L., Aras, E., Norrie, M.C.: Tilt-and-tap: framework to support motion-based web interaction techniques. In: Cimiano, P., Frasincar, F., Houben, G.-J., Schwabe, D. (eds.) ICWE 2015. LNCS, vol. 9114, pp. 565–582. Springer, Heidelberg (2015)
8. Husmann, M., Heyder, N., Norrie, M.C.: Is a framework enough? Cross-device testing and debugging. In: Proceedings of the EICS (2016)

Multi-device UI Development for Task-Continuous Cross-Channel Web Applications

Enes Yigitbas[1(✉)], Thomas Kern[2], Patrick Urban[2], and Stefan Sauer[1]

[1] S-lab - Software Quality Lab, Paderborn University, Zukunftsmeile 1,
33102 Paderborn, Germany
{eyigitbas,sauer}@s-lab.upb.de
[2] Wincor Nixdorf International GmbH, Heinz-Nixdorf-Ring 1,
33106 Paderborn, Germany
{thomas.kern,patrick.urban}@wincor-nixdorf.com

Abstract. The growing number of various types of web-enabled smart devices presents a special challenge for retail banks. In the world of Omni-Channel-Banking, customers demand a flexible and easy usage for carrying out their banking activities. Establishing such an Omni-Channel-Banking experience is a challenging task that requires support for the development of heterogeneous user interfaces (UIs) allowing flexible access to different channels (e.g. PC, Smartphone, ATM) and a seamless hand-over between these channels to allow task-continuity for the customer. Therefore, we present a model-based solution architecture for the development of multi-device UIs. Our solution architecture minimizes recurrent UI development efforts for different channels and enables data synchronization between them. To show the feasibility of our approach, we present an industrial case study, where we implement a cross-channel banking web-application that enables a modern customer experience.

Keywords: Model-based development · Multi-device UI development · Liquid software development · Cross-channel web applications · Self-service systems

1 Introduction

The growing number of various types of web-enabled smart devices (e.g. smartphones, smartwatches, tablets, etc.) presents a special challenge for retail banks. Customers demand a flexible and easy usage for carrying out their banking activities. While customers accessed banking services solely via isolated channels

This work is based on "KoMoS", a project of the "it's OWL" Leading-Edge Cluster, partially funded by the German Federal Ministry of Education and Research (BMBF).

S. Casteleyn et al. (Eds.): ICWE 2016 Workshops, LNCS 9881, pp. 114–127, 2016.
DOI: 10.1007/978-3-319-46963-8_10

(through banking personnel or ATM) in the past, using different channels during a transaction is nowadays increasingly gaining popularity. Depending on the situation, customers are able to access their banking services where, when and how it suits them best. In the world of Omni-Channel-Banking, customers are in control of the channels they wish to use, experiencing a self-determined "Omni-Channel-Journey". For example, if the customers pursue an "Omni-Channel-Journey" for a payment cashout process, they can begin an interaction using one channel (prepare cashout at desktop at home), modify the transaction on their way on a mobile channel, and finalize it at the automatic teller machine (ATM) (see Fig. 1). Thus, Omni-Channel-Banking brings the industry closer to the promise of true contextual banking in which financial services become seamlessly embedded into the lives of individual and business customers.

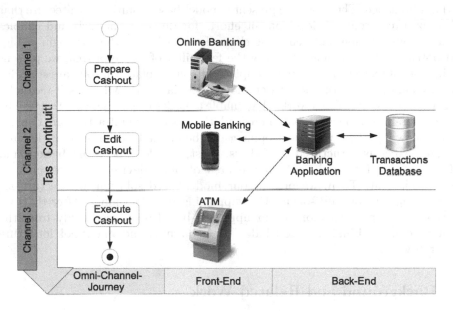

Fig. 1. Example scenario: omni-channel-journey with task-continuity

Table 1. Multi-channel-banking vs. omni-channel-banking

Multi-channel-banking	Omni-channel-banking
Fixed channel usage	Flexible channel usage
Separation of channels	Integration of channels
Data redundancy in channels	Data synchronization between channels
Little or no channel switch	Continuous channel switch

The advancement from Multi-Channel- to Omni-Channel-Banking (compare Table 1) is a difficult task for developers of such systems. Developers are facing the following challenges:

- **C1:** Support for heterogeneous user interfaces (UIs) allowing access for different channels (e.g. PC, Smartphone, ATM).
- **C2:** Support for a flexible channel usage depending on the context.
- **C3:** Support for a seamless handover between channels allowing task-continuity. When the user moves from one device to another, the user is able to seamlessly continue her task.

Tackling these issues by combining and integrating heterogeneous channels on a cross-platform software infrastructure imposes huge efforts for development and maintenance. Therefore, we present a model-based solution architecture that minimizes recurrent UI development efforts for different channels and enables data synchronization between these channels according to the paradigm of Liquid Software Development. To show the feasibility of our approach, we present an industrial case study, where we implement a cross-platform software stack for Omni-Channel cash transactions employing the latest HTML5-based technologies and open source frameworks that enable a modern customer experience. The implementation of our web-based cross-platform software infrastructure demonstrates how liquid software development can be applied to the banking domain.

The paper is structured as follows: First, we describe some background information and related work in the area of multi-device and cross-channel UI development. Then, we present our model-based solution architecture for the development of multi-device UIs supporting task-continuity. After that, we present the implementation of our approach based on a case study from the banking domain. Finally, we conclude with a summary and an outlook for future research work.

2   Background and Related Work

In recent years, a number of approaches have addressed the problem of interacting with liquid software applications distributed across various types of smart devices. Our work is inspired by and based on existing approaches from the area of distributed user interfaces (DUIs). In this section, we especially review prior work that explores the development of multi-device and cross-channel user interfaces (UIs) supporting task-continuity.

2.1   Multi-device UI Development

The development of multi-device UIs has been subject of extensive research [1] where different approaches were proposed to support efficient development of UIs for different target platforms. On the one hand, model-based UI development approaches were proposed which aim to create multi-device UIs based on the transformation of abstract user interface models to final user interfaces.

Two widely studied approaches are MARIA [2] and IFML[1] that support the abstract modeling of user interfaces and their transformation to multi-device UIs including web interfaces. In [3] we present a specialized approach for model-based development of heterogeneous UIs for different target platforms including self-service systems. On the other hand there are also existing approaches like Damask [4] and Gummy [5] following the WYSIWYG paradigm. While Damask is a prototyping tool for creating sketches of multi-device web interfaces, Gummy is a design environment for graphical UIs that allows designers to create interfaces for multiple devices using visual tools to automatically generate and maintain a platform-independent description of the UI. While above mentioned approaches support the development of multi-device UIs regarding specification and generation of UIs for different target platforms, they do not cover mechanisms to support channel switches and data synchronization between different target platforms at runtime.

2.2   Cross-Channel UI Development

Previous work by the research community has covered concepts and techniques to dynamically support the distribution of UIs by supporting task-continuity for the end-users. One of the concepts is called *UI migration*, which follows the idea of transferring a UI or parts of it from a source to a target device while enabling task-continuity through carrying the UI's state across devices. In [8], we present a model-based framework for the migration and adaptation of user interfaces across different devices. In [6], the authors present an agent-based solution to support migration of interactive applications among various devices, including digital TVs and mobile devices, allowing users to freely move around at home and outdoor. The aim is to provide users with a seamless and supportive environment for ubiquitous access in multi-device contexts of use. In the case of web applications, most solutions rely on HTML proxy-based techniques to dynamically push and pull UIs [7]. An extension of this concept is presented in [9], where the authors propose XDStudio to support interactive development of cross-device UIs. In addition, there is also existing work on the specification support for cross-device applications. In [10] for example, the authors present their framework Panelrama which is a web-based framework for the construction of applications using DUIs. In a similar work [11], the authors present Conductor, which is a prototype framework serving as an example for the construction of cross-device applications.

Leaning on the existing concepts of cross-channel UI development, we present a model-based solution architecture for multi-device UIs that supports task-continuity. By extending existing architectural cross-device application frameworks to the banking domain, we aim to support an omni-channel banking experience for the customers.

---

[1] http://www.ifml.org.

3   Solution Architecture

In this section, we present a model-based solution architecture for multi-device UI development in order to tackle the motivated challenges C1, C2 and C3. Figure 2 shows our solution architecture which is divided into three main steps: *Modeling*, *Transformation*, and *Execution*.

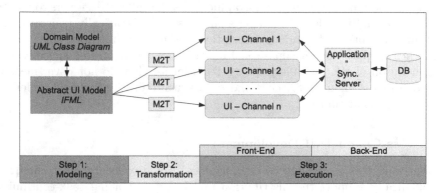

Fig. 2. Model-based solution architecture for multi-device UIs supporting task-continuity

For supporting the development of various UIs allowing access to the different channels and minimizing recurrent development efforts in establishing the needed *Front-Ends* (C1), we have a modeling step in our solution architecture. In the modeling step, a *Domain Model* described as UML class diagram and an *Abstract UI Model* based on the *Interaction Flow Modeling language (IFML)*, serve as specification of the data entities as well as structure, content and navigation needed to characterize the UI in an abstract manner. Based on the IFML *Abstract UI model*, which is referencing the *Domain Model*, a transformation is defined to generate heterogeneous UIs for the different *UI-Channels (1..n)*. Therefore, in the transformation step, several *model-to-text transformation (M2T)* templates are defined that transfer the *Abstract UI models* into the final UIs, running on different target platforms in order to support access to the different channels (Front-End). By generating different UI views and supporting different *UI-Channels*, the users are able to flexibly select the channel of their choice depending on the context (C2). In order to support a seamless handover between channels and allowing task-continuity for the user (C3), our solution architecture includes an *Application and Synchronization Server* in the *Back-End*, which is responsible for storing and sharing of data (e.g. UI state or user preferences). The UI state, including entered input data by the users, is stored and restored, allowing the user to move across channels while seamlessly continuing her task.

4   Instantiation of Development Process

In this section, the instantiation of the development process according to the previously described solution architecture is presented in more detail. To show the
feasibility of our approach, we first present the setting of an industrial case-study
dealing with the implementation of a cross-platform software infrastructure for
Omni-Channel cash transactions employing web-based technologies. After that,
we present the realization of the solution architecture by describing the implementation of the different steps.

4.1   Setting of the Case Study

Our "Omni-Channel-Banking" case-study supports a variety of different channels to access banking services. Figure 3 shows its overall architecture.

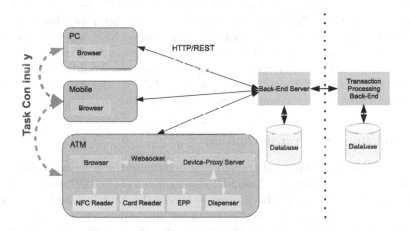

Fig. 3. Case-study application architecture

On each device - *PC, Mobile, ATM* - the client application is running as
a single-page web application inside a browser. The application communicates
with a *Back-End Server*, which is responsible for

– serving an application to the browser, adapted to a particular target device,
– serving application specific data to the client via HTTP/REST,
– managing application state and user preferences,
– requesting information from a *Transaction Processing Back-End* and serving
  it to the client,
– sending financial transactions to the *Transaction Processing Back-End* for
  execution.

The data format for all data exchanged through HTTP/REST requests is *JavaScript Object Notation* (JSON).

The *Transaction Processing Back-End* is not part of our application, but represents an existing infrastructure for processing financial transactions. The *Back-End Server* communicates with this transaction processing system. The communication protocol between the *Transaction Processing Back-End* and our sample application's *Back-End Server* depends on an existing infrastructure. Thus, the *Back-End Server* needs to provide a custom adapter for interfacing with this system.

In our case study, PC and mobile applications are identical concerning their functionality. The main difference comes from adaptation to different screen sizes and operation through a touch screen. This also includes spreading of functionality on the mobile device over multiple dialogs, compared to the PC application. In contrast to PC and mobile clients, the application architecture of the ATM client is significantly different. This is due to the need for supporting a whole variety of ATM specific hardware devices, like *NFC Reader*, *Card Reader*, *Encrypting Pin Pad (EPP)*, *Cash Dispenser*, etc. For interoperability reasons, ATM vendors are using a common software stack called XFS, which is layered on top of device specific drivers. XFS stands for *Extensions for Financial Services* and is standardized by CEN, the *European Commitee for Standardization*. Since a browser itself can not directly access the XFS-API, we delegate device control to a *Device-Proxy Server* running directly on the ATM.

4.2   Modeling and Transformation

For realizing the modeling and transformation step of the solution architecture, we have implemented a model-based UI development (MBUID) process which is depicted in Fig. 4. This MBUID process supports the modeling and transformation of the UIs, which are the view parts of the single-page application rendered as HTML5 by the browser. Using the open source IFML Editor Eclipse plugin[2], developers are able to specify the domain and abstract UI model. For transforming these models into final web UI views, we implemented an Xtend[3] plugin that maps the IFML model elements to specific HTML5 elements. The Xtend plugin includes different Xtend templates to transfer the IFML source model into web UIs supporting manifold devices.

During the transformation process, the application's view is built upon basic components with a custom look &feel, like buttons, text input fields, dropdown lists, tables, etc. As a basis for these components, we did not use AngularJS directives, but implemented components based on the *HTML5 Web Components*[4] specification promoted by Google as W3C standard.

---

[2] http://ifml.github.io.

[3] http://www.eclipse.org/xtend.

[4] https://www.w3.org/TR/components-intro.

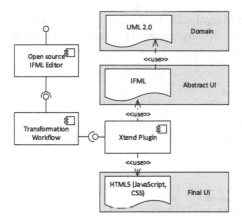

Fig. 4. Implemented model-based UI development process

Our custom components are sensitive to the application environment they are being used in (desktop, mobile, ATM) and adapt themselves accordingly. On mobile devices, for example, buttons are larger and more suitable for touch operation than on desktop devices.

Fig. 5. Buttons and text fields for desktop and mobile

Figure 5 shows buttons and text input fields. Their desktop representation is depicted on the left side, their mobile appearance on the right side of the picture.

During the transformation process for all device classes, a button is created the same way:

```
<komos-button colorscheme="cs1" ng-click="confirm()">
  Confirm
<komos-button>
```

The following example shows how to create a text input field with a label by mapping an IFML simple field element to the following code snippet:

```
<komos-textfield label="Current_PIN" ng-model="model.currentPin">
  </komos-textfield>
```

In order to provide a unified layout management for our application, our model-to-text (M2T) transformation process implements a custom layout manager. It provides an easy to use grid layout system, based on row and column elements realized as AngularJS directives. Under the hood, it uses HTML5 Flexbox. The following listing shows the generated code snippet to create the dialog shown in Fig. 6.

```
<komos−container>
  <komos−row>
    <komos−column span−3>
      <komos−label>Username</komos−label>
    </komos−column>
    <komos−column span−8>
      <komos−textfield name="username" ng−model="model.username">
        </komos−textfield>
    </komos−column>
  </komos−row>

  <komos−row>
    <komos−column span−3>
      <komos−label>Password</komos−label>
    </komos−column>
    <komos−column span−8>
      <komos−textfield name="password" ng−model="model.password">
        </komos−textfield>
    </komos−column>
  </komos−row>

  <komos−row>
    <komos−column offset−3 span−8>
      <komos−button colorscheme="cs1" ng−click="login(form)">
      Login
      </komos−button>
      <komos−button colorscheme="cs5" ui−sref="public.signup">
        Register
      </komos−button>
      <komos−button colorscheme="cs3" ng−click="reset(form)">
        Cancel
      </komos−button>
    </komos−column>
  </komos−row>
</komos−container>
```

Fig. 6. Login dialog

4.3   Execution (Front-End)

While the previous subsection presented our MBUID process to support the modeling and generation of view aspects of the *Front-End*, this section deals with the execution step. In this context, we especially present the controller part of the *Front-End*, which is responsible for application logic and communication with the *Back-End Server*. In conjunction with this topic, we also present the aspect of channel handover and task-continuity.

As shown in Fig. 7, the *Front-End* consists of a HTML5/JavaScript single-page application running in a web browser. It exchanges JSON messages with the *Back-End Server* through HTTP/REST.

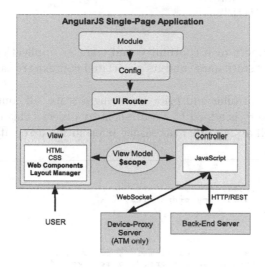

Fig. 7. Front-end architecture

The browser application's main building blocks are:

– AngularJS[5]: Google's open-source web application framework for developing single-page applications in JavaScript
– UI Router[6]: flexible client-side routing with nested views in AngularJS
– Web Components: UI components with custom look & feel
– Layout Manager: custom layout manager

AngularJS supports the model-view-controller (MVC) design pattern by decoupling the application's presentation layer, which is defined through HTML5 (see previous subsection), from the model and application logic by two-way data-binding through a $scope object. In addition, AngularJS provides a variety of other services, including modularization and definition of custom directives.

UI Router is the client-side routing component of AngularJS and the central key component to implement task continuity. The developer assigns a particular application state, identified by a name (protected.main), with a view (main.html) and a controller (MainCtrl):

---

[5] https://angularjs.org.
[6] https://github.com/angular-ui/ui-router/wiki.

```
angular.module('komosApp').config(function ($stateProvider) {
  $stateProvider
    .state('protected.main', {
      url: '/',
      templateUrl: 'protected/main/main.html',
      controller: 'MainCtrl',
      authenticate: true
    });
});
```

In order to support task continuity and transfer application state between devices, the current state name and its associated context are saved to the *Back-End Server*.

Inside a view controller and prior to saving a state, all context information necessary for recovery is added to a state-context object. This includes the UI's view-model, as well as any other necessary information associated with the current state.

```
var context = {
  // the view model:
  viewModel: $scope.model,
  // state specific arbitrary properties:
  param1: someValue,
  data: someData
};

PersistStateService.save('protected.main', context, function (err, data) {
  if (err) model.errors.message = err.data.message;
});
```

We implemented an AngularJS service named `PersistStateService`, which converts the object `context` to JSON and sends it to the *Back-End Server*, where it is stored under the name of the state, e.g. `protected.main`. To invoke a previously saved state, the application just needs to retrieve the current state name and invoke it:

```
$rootScope.$state.go('protected.main');
```

On instantiation of the AngularJS controller associated with this state, the controller calls the service's restore method to retrieve the previously stored information:

```
PersistStateService.restore('protected.main', function (err, context) {
  if (err) {
    model.errors.message = err.data.message;
  } else {
    // context now contains the previously saved information
    $scope.model = context.viewModel; // this updates the UI!!
    someValue = context.param1;
    someData = context.data;
  }
});
```

Both, saving and retrieving context data for a state happens within the same controller. Each controller knows exactly which data needs to be saved in order to be able to restore itself. This information is hidden from other parts of the application. The only knowledge necessary from the outside is the name of the state, `protected.main` in our example.

Because of AngularJS' two-way data-binding, assigning the view-model to `$scope.model` immediately updates the view.

4.4   Execution (Back-End)

The application's *Back-End* is implemented in JavaScript (see Fig. 8) and uses Node.js[7] as its runtime environment. It is built upon Google's V8 JavaScript engine also used by Google Chrome and provides a high-performance runtime environment for non-blocking and event-driven programming.

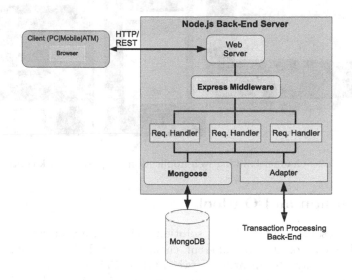

Fig. 8. Back-end architecture

ExpressJS[8], which is a middleware for Node.js, provides components for processing of requests and routing. An application sets up request handlers, which are automatically invoked when a client request arrives. Within a request handler, the request is processed, a response is prepared and returned. Request handlers communicate with the database or *Transaction Processing Back-End*.

The document database *MongoDB*[9] belongs into the category of NoSQL ("Not Only SQL") databases. In this context, a "document" consists of a user-defined data structure of key-value pairs, which is associated with a key. Documents can also contain other documents. The schema of a database is dynamic and can be modified at runtime. To access the database in an object-oriented fashion, we use an *Object Document Mapper* called *Mongoose* on top of MongoDB's Node.js driver.

The instantiation of our solution architecture and interaction of all described technologies resulted in the demonstrator which is shown in Fig. 9. Our demonstrator shows the implemented cross-channel web-application that supports different channels (Desktop, Tablet, and ATM) for a cash payout process enabling task-continuity for the customers.

---

[7] https://nodejs.org.

[8] https://expressjs.com.

[9] https://www.mongodb.org.

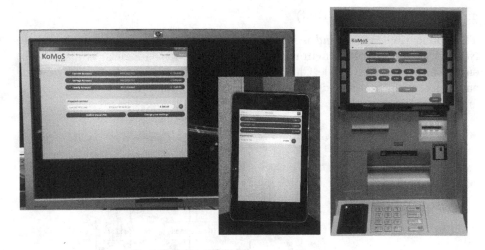

Fig. 9. Cross-channel banking web application supporting task-continuity

5   Conclusion and Outlook

This paper presents a model-based solution architecture that supports the efficient development of UIs for different channels (e.g. PC, Smartphone, ATM) and enables data synchronization between them. This solution offers end-users a flexible and easy usage for accessing their services through variable channels and a seamless hand-over between channels allowing task-continuity. We showed the feasibility of our approach based on a cross-channel banking web-application that was implemented according to our solution architecture. The implementation of our case study includes a cross-platform software infrastructure for Omni-Channel cash transactions employing the latest web technologies and open source frameworks. In ongoing work we are developing and extending the model-to-text transformation process in order to support the generation of dynamic UI aspects (e.g. input validation, controller artifacts, etc.). Our future work will focus on studies with web designers and developers to further evaluate the efficiency and effectiveness of our approach.

References

1. Paternò, F., Santoro, C.: A logical framework for multi-device user interfaces. In: Proceedings of the 4th ACM SIGCHI Symposium on Engineering Interactive Computing Systems (EICS 2012), pp. 45–50. ACM, New York (2012)
2. Paternò, F., Santoro, C., Spano, L.D.: MARIA: a universal, declarative, multiple abstraction-level language for service-oriented applications in ubiquitous environments. ACM Trans. Comput.-Hum. Interact (2009)
3. Yigitbas, E., Fischer, H., Kern, T., Paelke, V.: Model-based development of adaptive UIs for multi-channel self-service systems. In: Sauer, S., Bogdan, C., Forbrig, P., Bernhaupt, R., Winckler, M. (eds.) HCSE 2014. LNCS, vol. 8742, pp. 267–274. Springer, Heidelberg (2014)

4. Lin, J., Landay, J.A.: Employing patterns and layers for early-stage design and prototyping of cross-device user interfaces. In: Proceedings of the SIGCHI Conference on Human Factors in Computing Systems (CHI 2008), pp. 1313–1322. ACM, New York (2008)

5. Meskens, J., Vermeulen, J., Luyten, K., Coninx, K.: Gummy for multi-platform user interface designs: shape me, multiply me, fix me, use me. In: Proceedings of the Working Conference on Advanced Visual Interfaces (AVI 2008), pp. 233–240. ACM, New York (2008)

6. Paternò, F., Santoro, C., Scorcia, A.: Ambient intelligence for supporting task continuity across multiple devices and implementation languages. Comput. J. **53**(8), 1210–1228 (2010)

7. Ghiani, G., Paternò, F., Santoro, C.: Push and pull of web user interfaces in multi-device environments. In: Proceedings of the International Working Conference on Advanced Visual Interfaces (AVI 2012), pp. 10–17. ACM, New York (2012)

8. Yigitbas, E., Sauer, S., Engels, G.: A model-based framework for multi-adaptive migratory user interfaces. In: Kurosu, M. (ed.) Human-Computer Interaction. LNCS, vol. 9170, pp. 563–572. Springer, Heidelberg (2015)

9. Nebeling, M., Mintsi, T., Husmann, M., Norrie, M.: Interactive development of cross-device user interfaces. In: Proceedings of the SIGCHI Conference on Human Factors in Computing Systems (CHI 2014) (2014)

10. Yang, J., Wigdor, D.: Panelrama: enabling easy specification of cross-device web applications. In: Proceedings of the SIGCHI Conference on Human Factors in Computing Systems (CHI 2014), pp. 2783–2792. ACM, New York (2014)

11. Hamilton, P., Wigdor, D.J.: Conductor: enabling and understanding cross-device interaction. In: Proceedings of the SIGCHI Conference on Human Factors in Computing Systems (CHI 2014), pp. 2773–2782. ACM, New York (2014)

Liquid Context: Migrating the Users' Context Across Devices

Javier Berrocal[1]([⊠]), Jose Garcia-Alonso[1], Carlos Canal[2], and Juan M. Murillo[1]

[1] University of Extremadura, Cáceres, Spain
{jberolm,jgaralo,juanmamu}@unex.es
[2] University of Málaga, Málaga, Spain
canal@lcc.uma.es

Abstract. The ever increasing computing and storage capacity of smart devices are enabling users to perform in them tasks that until now were relegated only to devices with high computing capabilities (such as PCs or laptops). Empowering users to employ in each moment the device that best adapts to each concrete situation. This demands that the applications deployed on them should provide a consistent user experience when users migrate from one device to another. The Liquid Software paradigm facilitates the development of this kind of applications. However, in order to get a more satisfying user experience, these applications should also be adaptable to the specific context of each user. This position paper presents the concept of Liquid Context, being the contextual information that migrates across devices along with the applications and their data. In addition, we also propose an architecture for the development of context-aware liquid applications. These techniques will improve the usability and the user experience of liquid applications.

Keywords: Liquid software · Context-aware · Liquid context · Virtual profile

1 Introduction

The ever increasing computing and storage capacity of mobile devices has led to a change in how users behave with these devices [16,28]. Currently, many of the activities that were only performed on PCs (such as writing documents, or editing photos or videos) are also done in any of these devices. In addition, nowadays, other devices, such as televisions, cars or watches, are getting smarter. So that, it is expected that they will take a more prominent role for executing activities that today are relegated to a PC or a laptop [35], further increasing the diversity of devices that we can use. Thus, which device to use will depend on the context and the preferences of each user in each moment. Even, a user may want to start an activity in a particular device and, when her situation or context changes, to seamlessly roam to another device to continue it. In this scenario, users will demand a similar usage experience on every device.

© Springer International Publishing AG 2016
S. Casteleyn et al. (Eds.): ICWE 2016 Workshops, LNCS 9881, pp. 128–141, 2016.
DOI: 10.1007/978-3-319-46963-8_11

Currently, migrating the activities that are being done in an application from one device to another requires a great effort [6]. Users have to install the application on each device, to synchronize the data used on each of them and, for each change, to manually configure the instance of the application to its state in the previous device. For example, if a user is using a music player on a PC and wants to continue the music session on her smartphone, although the application and all the songs could be already synchronized between devices, she must select again the list and the specific song that she was listening. This entails a repetitive effort for the user that could be reduced if the application status would also be synchronized.

In the last few years, researches have been working on the *Liquid Software* paradigm [33]. This paradigm proposes an architectural style and a range of technologies to develop applications that can migrate its business logic, its data and the application state to different devices. This provides a seamlessly user experience when they roam across devices [24]. Thus, if the music player were developed following this style, the user would be able to change of device without having to select again the list nor the song.

However, to develop systems that can migrate their logic, data, and state is not enough for some applications. Currently, Internet of Things (IoT) [9] and Web of Things (WoT) [11] applications interact with a huge amount of devices. For instance, a music player based on the WoT paradigm would interact with the hi-fi system, the television or the speakers. In order to increase the user experience, these interactions and, even, the system behaviour may depend on the users' preferences and context. So far, these preferences and context were manually set by the user. With the aim of reducing the effort required to configure them, different approaches, such as Ambient Intelligence [5,22] and Context-Aware [2,21], have been defined to identify the users' needs and preferences and to develop applications that can adapt their behaviour to them [14].

The need of developing applications aware of their users' context will also be a requirement for liquid applications. These applications, when they migrate from one device to another, should also be aware of what are the user's needs and preferences. So that, they can dynamically adapt themselves to her profile in order to fully provide a transparent roaming. This implies that the user's virtual profile must also be migrated, along with the application logic, data and state, from one device to another. For example, the music player could suggest songs depending on the user's preferences and mood. But, for this functionality to be correctly executed on any device, the user's profile must be available and updated at any device at which the user roam.

In this position paper the authors outline the concept of Liquid Context, as the ability for the virtual profile to migrate across devices depending on the specific requirements of the applications deployed on them. In addition, this concept is added as an enrichment of the architecture detailed by Mikkonen et al. for Liquid Web Applications [24]. To that end, some modules responsible for building and managing the user's virtual profile is added to the architecture. These modules allow the development of context-aware liquid applications that can adapt themselves to the user's preferences independently of the device in which they are deployed. Liquid Context is part of a more extensive research work, called Situational-Context [3], improving the integration and the interactions between people and smart things in multi device systems.

To deeply detail the concept of Liquid Context, the rest of the paper is structured as follows. Section 2 presents the motivations for the development of Liquid Context. Section 3 defines the concept of Liquid Context, the main challenges that should be faced during its implementation and a proposed architecture for context-ware liquid applications. Section 4 contains some related works. And, Sect. 5 details the conclusions and the future works.

2 Motivations

In order to better illustrate the motivation of this work, the following running example is used. Petter is an accountant. Today, he has a reservation in a new restaurant that recently opened in the city. Before leaving his workplace, in the PC, he starts a navigation application in order to review the route that he has to follow to reach the restaurant. Then, he leaves work and gets into his car to head to the restaurant. The ideal behaviour of these devices would be that, when Petter gets into his car, the route viewed in the PC is automatically migrated from the PC to the car. With current applications, Petter would have to start the navigation application in the car, go to the history, and select the last viewed route. This leads to a waste of time and increases his frustration.

One of the central aspect of Liquid Software [33] is to be able to fluidly move an application, and its associated data, from one device to another without requiring users to have an active attention in the roaming, or without having to remember complex steps [34]. Thus, if the application used by Petter was developed following the key requirements of Liquid Applications, its logic, the selected route and the state of the application (for instance, the concrete part of the route that Petter was viewing) would be transparently migrated from the PC to the car.

Current applications, in order to provide a greater usability and user experience, can adapt their behaviour on the users' contextual information [1]. Nowadays, there are a lot of works focused on gathering the user's contextual information and performing high level inferences in order to create more comprehensive virtual profiles [8,27]. Likewise, there are some works focused on defining new programming languages or new software adaptation techniques that are able to take into account the virtual profile of the users [13,14,20]. The applications developed with these techniques can adapt their behaviour to the needs and the situation of each user. Thus, in the running example, if the navigation application were develop to be context-aware, when Petter set the destination place and the route is calculated, the application would query the virtual profile in order to know if Petter usually goes to any other places before lunch. The virtual profile could have stored that he usually goes to his wife's workplace in order to pick her up and have lunch with her. Therefore, the first route that the system would offer to Petter would be going through the workplace of his wife, improving the Petter's experience and satisfaction.

With the aim of providing maximum value, liquid applications should also adapt their behaviour to the user's profile. In the running example, if Petter's profile is also migrated to the car navigator, any recalculation of the route (due to traffic jams or incidents in the road) would also take into account that he should go first to the workplace of this wife.

In order to develop context-aware liquid application, the virtual profiles could be integrated together with the domain data of the application, so that they could be migrated altogether. However, their nature is quite different. First, the application data are generally used only by that application, while the virtual profile is generic and can be used by any application. Second, the construction of the virtual profile requires the execution of inference rules that need high computing capabilities that not all devices have. Finally, virtual profiles have a lot of information, but applications normally only use a part of it, so a selective migration could be done in order to increase the data transfer efficiency. Therefore, specific techniques to efficiently migrate the virtual profiles are needed.

Currently, there are different approaches, such as [19,30], storing the users' virtual profiles in a central server. So that, they are always available to be accessed by any device. Nevertheless, this also requires a constant communication with the sever, both to store the new contextual information in the users profile and to access to them. This, inevitably incurs high communication overhead [12].

In order to move data closer to mobile users, some studies propose to distributively store them on different mobile devices or on the users' main device. In the Ambient Intelligence paradigm, for example, some works use multi-agents systems to gather the users' context and to react in a proactive and autonomously way to it [26,31,32]. Even, the authors of this paper proposed, in previous works, to use the smartphones as the key element to create and maintain these profiles [10,25]. However, the constant access to virtual profiles stored on mobile devices can lead to a quickly drain of their battery, which is a crucial aspect for them and for the success of mobile applications [23,29]. Therefore, new methods for transferring and migrating the virtual profiles from one device to another are needed. This requires that when an application is migrated, along with its logic, data and state, the user's virtual profile and the inference capabilities (i.e., the inference engine and the rules) should also be migrated from one device to the other.

In this paper the concept of Liquid Context is detailed as the contextual information that can migrate or flow from one device to another. Moreover, an architecture is proposed for the development of context-aware liquid applications.

3   Context-Aware Liquid Applications

3.1   Liquid Context

This section contains some important concepts to explain what the Liquid Context is and what its main requirements are. As defined in [1] by Abowd et al., we understand that the *context* is "any information that can be used to characterize the situation of an entity. An entity is a person, place, or object that is considered relevant to the interaction between a user and an application, including the user and applications themselves". All the contextual information of an entity is encapsulated into a virtual profile, representing that entity in the connected world.

Concretely, the *virtual profile of a person* contains all the data gathered by the sensors associated to her devices, both internal and external, and all the actions performed with them. This profile not only maintains the fresh and updated data collected in the latest instant, but it keeps a history with all the gathered data. This history is ordered according to the date and time at which the data were collected, forming a timeline with all the contextual information. All these data are further processed to make inferences and draw high-level information (such as the preferences, activities or profession of a person). The virtual profile of a user is divided into the following parts:

- A *Basic Profile* containing the dated raw information of the person's context and the relationships with other people and devices. This profile can be seen as a timeline with the changes and interactions that happened to her.
- *Social Profile*. This profile contains the results of the high level inferences performed over the Basic Profile.
- *Goals* detailing the status of the environment the person desires to reach regarding different issues of the daily life, ranging from temperature to social interactions or the traffic conditions. These Goals can also be inferred from the Basic and Social Profiles.

For the running example, the basic profile of Petter would contain information on his locations and the people and devices with which he interacts. This information, being temporally ordered, could be processed to infer where his workplace is, who his wife is, what the places he more frequently visits are or what the routes he usually follows in his car journeys are. All this information would be stored in his social profile. In addition, both profiles could be further analysed in order to identify Petter's Goals, such as that he usually desires driving on low congested roads.

The virtual profile could be stored in the user's mainly device, which is her companion device, but the information it contains would be required by all devices and applications adapting their behaviour to her preferences. The *Liquid Context* is the ability of all or part of the virtual profile of a person to migrate from one device to another without having her to take any action, nor having to manage the information flow. This allows a user, when changing from one device to another, to have her virtual profile automatically available on the new device to be used.

In the running example, the navigation application would require to be liquid at least the part of the profile indicating that Petter picks his wife up every day to go to lunch, so that it could be transferred from the PC to the car.

3.2   Challenges

The development of Liquid Context and how it should be efficiently transferred across devices presents a set of challenges. The main challenges to make the people virtual profile liquid are:

Virtual profile location. The liquid context should always be available to be consulted by any device when an application migrates from one device to another. To achieve this, the virtual profile could be stored on a central server or in the final devices. Being stored in a central server, it is more easily accessible. However, this also requires all sensors and devices to be constantly sending updated information to the server to maintain it. In addition of requiring all devices to constantly have global connectivity independently of the ambient conditions, which leads to an increase in the data traffic and in the battery usage, which may jeopardize the success of any application.

If the virtual profile is kept on the user's companion device, all the data gathered by its sensors would be locally stored, reducing the communication requirements and the battery consumption. Nevertheless, this also implies that when a user roams from once device to another, her virtual profile should also be migrated. The size of the transferred information and the frequency at which a user changes of device will indicate which alternative is the most efficient one.

Information transferred. Making the complete virtual profile liquid allows any device to perform any operation based on the contextual information. But, this also implies that a huge amount of information should be migrated. The virtual profile stores a history with all the data gathered by the different sensors and all the information inferred. Depending on the number of sensors and the frequency at which they gather data, this history can be quite large.

Nevertheless, applications normally do not need access to the full virtual profile. Depending on the purpose of the applications, they would only require the access to a subset of the data. Therefore, if an application could define the subset of contextual information required for its proper execution, only that subset would be migrated, consequently the size of the transferred information would be reduced. The virtual profile would be migrated, thus, on demand depending on the applications needs.

Summarizing information. In order to further reduce the transferred information, the subset of data required by an application could be summarized. For the running example, the navigation application could require information on the most frequented places. But, for the specific usage of guiding Petter to the restaurant, only the most frequented places at noon would be necessary. Techniques for summarizing the contextual information according to the specific situation of usage of the application are also needed.

Simultaneous multi-device usage. Liquid applications, following the Liquid Manifesto, can not only be executed sequentially on different devices, but they can also be executed simultaneously on multiple devices. For example, Petter could be simultaneous using the navigation application both in the car navigator and in his phone. In this situation, the same contextual information would be duplicated in all devices. This implies that any update performed on the profile stored in a device has to be duplicated in the other devices in order to maintain the data synchronized.

Migrating the inference engine. Inference engines are normally used to build the Sociological profile and to deduct the user's goals. Normally, this engines have to be executed in devices with high computation capabilities. Therefore, when the virtual profile is migrated, it should also be identified whether the new device has enough capabilities to run it. If it does not have them, a strategy should be defined to identify where to deploy the engine and how to maintain updated the profiles with its results.

Liquid Context and IoT. Even if a user does not directly use multiple devices simultaneously with a single application, due to the proliferation of smart devices connected to the Internet and the IoT, there may be several devices requiring the context of a person to adapt their behaviour. For example, in a smart car, the air conditioning would need to know the comfort temperature desired by the user and the radio would require his musical style. The liquid context should also allow the contextual information to flow to all devices requiring it to meet the user's goals. The part of the virtual profile that must be migrated to each device should depend on the capabilities that they have to meet the user's goals.

Predictability and degree of pro-activity. Interactions between devices must be predictable. That is, they should be triggered according to the gathered contextual information. In addition, users must be empowered to indicate the expected behaviour depending on each specific context, must be able to stop an interaction immediately and devices should also be able to detect when an interaction causes a rejection in the users. Thus, systems would have the pro-activity degree desired by their users.

Turning off the context monitoring. Finally, technology should provide users the ability to turn off any monitoring system and to stop registering the context and the activities done by them. This would allow users to temporarily leave the technological experience.

3.3   Architecture

Liquid Context could be incorporated to liquid software in order to develop Ambient Intelligent and context-aware liquid applications. Figure 1 shows a possible architecture for this systems. It is derived from the architecture for liquid applications defined in [24]. This architecture is based on the client-server architectural style. Of course, depending on the computing capacity of the targeted devices and the features of the applications, this architecture could also be designed using other architectural styles.

The proposed architecture is divided into a server and a client side. The server-side is responsible for performing the functionalities of the application that are independent of the user or that require a high level processing capacity.

The client-side, is responsible for interacting with the user and gathering the contextual information required to build the virtual profile.

The server-side contains part of the business logic and the application data. These should be generic parts of the liquid application that are not dependent nor the device nor the user's context. Thus, the user' context does not have to be stored on the server and can be maintained entirely on the client-side, reducing the amount of information transferred. Coming back to the running example, the different maps used by the navigation application and the algorithms for calculating the routes would be located on the server side of the application. Both elements are independent of the device and the user executing the application.

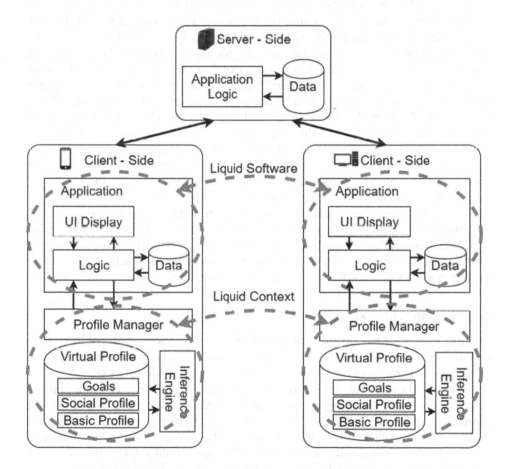

Fig. 1. Contex-aware liquid applications architecture.

The client-side contains, first, the user interface, adaptable to the different devices on which the application can be deployed. Second, the specific data of the application for that user. For instance, the colours or fonts preferred by Petter, or the route that he is currently following to reach the restaurant. And, third, the logic of the application dependent of the device and the user's context. The logic dependent of the device adapts the user interface and the behaviour of the system to the concrete device on which it is deployed. The logic dependent of the user's context accesses to the virtual profile to adapt the behaviour of the system. For the running example, this logic would be composed by the required algorithms for consulting Petter's profile and getting information on the places he usually visits at a specific hour. All these components should be developed following the principles defined in Liquid Software manifesto, so that they can be migrated from one device to another.

The client-side also contains the virtual profile of the user, containing the basic profile, the social profile and the goals; the inference engine, which includes all the inference rules for deducting the social profile and the goals; and the Profile Manager module, managing the migration of the Liquid Context. The Profile Manager, therefore, is the module responsible for minimizing the information transferred. It identifies whether the full virtual profile and the inference engine must be migrated to the new device, if it is enough transferring only a subset of the contextual information and if that subset can be further summarized. For this, the application should explicitly indicate which contextual information it requires to correctly continue the execution of the system on another device. For the running example, the navigation application would specify that information on the user's location and the frequency at which he visits every place would be needed. Thus, only this information would be transferred to the PC or the car navigator, saving data traffic and battery.

In addition, the Profile Manager module is responsible for maintaining synchronized the virtual profiles deployed on different devices. Thus, if the navigation application is simultaneously used in the car navigator and in the smartphone, the Profile Manager detects it and maintains both profiles constantly synchronized, so that both devices would have access to updated information.

3.4   Liquid Context and IoT

Finally, when an application is deployed on a device, the system could provide specific skills or capabilities to the device. Thus, if a navigation application is deployed on a smartphone, it acquires capabilities to guide its owner.

These skills allow the device to make decisions or to perform actions capable of modifying the environment and aimed at achieving the user Goals. In the running example, one of the goals inferred from Petter's context could be that he desires to "avoid highly congested roads". The navigation application, since it provides specific capabilities to achieve this goal, could get the virtual profile of Petter and compute it in order to define a strategy for calculating routes avoiding this kind of roads.

Like before, to make these decisions, a part of the virtual profile should be migrated to the smartphone. To be able to transfer only a part of the profile, the application should clearly define the skills that it provides to the device. Then, when the application roam across devices, the Profile Manager module computes these skills and the profile in order to identify the specific contextual information that should be migrated. In the running example, only the part of the profile related with traffic and roads information would be migrated. This would reduce the data traffic consumption.

Being able to migrate the user's profile to devices, in order to achieve the goals defined on that profile, reduces the effort required to configure and to interact with IoT systems. These systems are composed by a huge amount of devices, each one having different skills. The Liquid Context could be used to identify the specific part of the context that should be migrated to each device, reducing the data traffic and the battery consumption.

4 Related Works

Currently, there is a large amount of works related with the Liquid Context in the Context-Aware [15], Ubicomp [4], User Modelling [17] and Ambient Intelligence [22] areas. Some of their results will be used in the proposal detailed in this paper to address the different challenges identified. Below some of them are described.

There are works focused on managing people contextual information with the aim of improving their experience in multi-device systems [7]. An important subset of these approaches are working on storing the contextual information in central servers. Thus, the virtual profile are accessible from every device.

In [30] the authors indicate that ubiquitous computing advocates the construction of massively distributed systems that can be deployed in heterogeneous devices. These systems should take into account the user's context, adapting themselves to different situations. Therefore, they propose a middleware facilitating the development of context-aware agents. This middleware consists of an ontology describing the structure of the context to store, context providers gathering and sensing information, context synthesizer getting the information from the context provider to make high level information deduction, and context consumers – or applications. The aggregated contextual information is stored on specific agents or on central servers. This facilitates clients to access it. The authors also claim that this allows heterogeneous agents to seamlessly interact among each other.

In [18,19] the authors indicate that in order to enable context-awareness for distributed applications, new architectural styles are needed to support the gathering and interpretation of disseminated context attributes. Therefore, they propose a context-aware middleware based on the client-server architecture. In this middleware, a central server is responsible for structuring, representing, storing and interpreting the contextual information. The clients, which are mobile devices, are responsible for acquiring the context, sending it to the server, getting the processed information, and, by means of a specific API, working as a proxy for multiple applications. The contextual-aware applications deployed on these mobile devices, are responsible for adapting their behaviour to the contextual information.

Having the profiles in a central server requires a high communication between the devices and the server, both for storing new information and for consulting the virtual profile [12]. Thus, other works are focused on storing these profiles on devices closer to the users.

In [26] the authors focused on the Ambient Intelligence environment. They indicate that this environment covers a large number of devices and people. So, in order to achieve a better scalability, they have defined a generic middleware layer for transferring contextual information between devices. This middleware is based on software agents in which context-awareness is implemented both in the agent and in the logical topology of the agent system. Each agent in the middleware runs on a specific device. Therefore, a high decentralization of the contextual information is achieved. Each agent in the multi-agent system naturally accesses, processes and shares context information. The agents communicate between each other in order to exchange specific information and create a more complete context graph.

In addition, the authors of this work have proposed different approaches in this sense. In [10], the People as a Service (PeaaS) mobile-centric computing model is proposed. This model allows sociological profiles of people to be generated, kept, and securely provided to third parties as a service. The main idea behind PeaaS is that smartphones can be used to gather and create the virtual profile of their owners. These profiles are then provided as a service, in the same way that servers are used, for being consumed by other entities requiring the contextual information.

In [25], the PeaaS paradigm is used to facilitate the integration of IoT systems into the smartphone owners' life. To that end, smartphones are made the architectural core of these systems, using the IoT sensors to create richer sociological profiles of the phones' owners and using this information to better adapt the systems to their needs.

For accessing specific contextual information, these methods are adequate. However, if an application needs to have a more direct access to the virtual profile, then the required communication and the waste of battery associated to it could be too high. Therefore, methods, like the one proposed in this paper, are needed to make the virtual profile of a person liquid and to be able to migrate it from one device to another.

5   Conclusions and Future Work

In the near future, users will be connected to several heterogeneous devices and they will be able to deploy virtually any application on any device. For providing an engaging user experience, they will have to be perfectly adapted to the users' needs and preferences, independently the device on which each application is executed in each moment.

This position paper has presented the concept of Liquid Context. This contextual information is the part of the virtual profile of a user that can migrate from one device to another. This will allow user to roam across devices transparently and will reduce the effort required to reconfigure an application at each change.

We are currently working on formalizing the liquid context and on evaluating the consumption required to migrate it. In the next year, we will develop the solutions and the technology associated to the Liquid Context concept, ranging from technologies for migrating it to programming models for developing applications managing this context.

Acknowledgments. This work was partially supported by the Spanish Ministry of Science and Innovation (projects TIN2014-53986-REDT, TIN2015-67083-R and TIN2015-69957-R), by the Department of Economy and Infrastructure of the Government of Extremadura (GR15098), and by the European Regional Development Fund.

References

1. Abowd, G.D., Dey, A.K.: Towards a better understanding of context and context-awareness. In: Gellersen, H.-W. (ed.) HUC 1999. LNCS, vol. 1707, pp. 304–307. Springer, Heidelberg (1999)
2. Bellavista, P., Corradi, A., Fanelli, M., Foschini, L.: A survey of context data distribution for mobile ubiquitous systems. ACM Comput. Surv. **44**(4), 1–45 (2012)
3. Berrocal, J., Garcia-Alonso, J., Canal, C., Murillo, J.M.: Situational-context: a unified view of everything involved at a particular situation. In: Bozzon, A., Cudré-Mauroux, P., Pautasso, C. (eds.) ICWE 2016. LNCS, vol. 9671, pp. 476–483. Springer, Heidelberg (2016). doi:10.1007/978-3-319-38791-8_34
4. Caceres, R., Friday, A.: Ubicomp systems at 20: progress, opportunities, and challenges. IEEE Pervasive Comput. **1**, 14–21 (2011)
5. Cook, D.J., Augusto, J.C., Jakkula, V.R.: Ambient intelligence: technologies, applications, and opportunities. Pervasive Mob. Comput. **5**(4), 277–298 (2009)
6. Dearman, D., Pierce, J.S.: It's on my other computer!: computing with multiple devices. In: Proceedings of the SIGCHI Conference on Human Factors in Computing Systems. CHI 2008, pp. 767–776. ACM, New York (2008). http://doi.acm.org/10.1145/1357054.1357177
7. Denis, C., Karsenty, L.: Inter-usability of multi-device systems: a conceptual framework. In: Multiple User Interfaces: Cross-Platform Applications and Context-Aware Interfaces, pp. 373–384 (2004)
8. Gronli, T.M., Ghinea, G., Younas, M.: Context-aware and automatic configuration of mobile devices in cloud-enabled ubiquitous computing. Pers. Ubiquit. Comput. **18**(4), 883–894 (2014)
9. Gubbi, J., Buyya, R., Marusic, S., Palaniswami, M.: Internet of things (IoT): a vision, architectural elements, and future directions. Future Gener. Comput. Syst. **29**(7), 1645–1660 (2013)
10. Guillen, J., Miranda, J., Berrocal, J., Garcia-Alonso, J., Murillo, J.M., Canal, C.: People as a service: a mobile-centric model for providing collective sociological profiles. IEEE Softw. **31**(2), 48–53 (2014)
11. Guinard, D., Trifa, V., Mattern, F., Wilde, E.: From the internet of things to the web of things: resource-oriented architecture and best practices. In: Uckelmann, D., Harrison, M., Michahelles, F. (eds.) Architecting the Internet of Things, pp. 97–129. Springer, Heidelberg (2011)
12. Han, D., Yan, Y., Shu, T.: Context-aware distributed storage in mobile cloud computing. In: 2015 IEEE 12th International Conference on Mobile Ad Hoc and Sensor Systems (MASS), pp. 460–461, October 2015

13. Heo, S., Woo, S., Im, J., Kim, D.: IoT-MAP: IoT mashup application platform for the flexible IoT ecosystem. In: International Conference on the Internet of Things, pp. 163–170. IEEE (2015)

14. Hirschfeld, R., Costanza, P., Nierstrasz, O.: Context-oriented programming. J. Object Technol. **7**(3), 125–151 (2008). ETH Zurich

15. Hong, J.Y., Suh, E., Kim, S.J.: Context-aware systems: a literature review and classification. Exp. Syst. App. **36**(4), 8509–8522 (2009)

16. International Data Corporation (IDC): Mobile device users/non-users: print, scan, document management, worldwide (2015)

17. Kobsa, A.: Generic user modeling systems. User Model. User-Adap. Inter. **11**(1–2), 49–63 (2001)

18. Löwe, R., Mandl, P., Weber, M.: Context directory: a context-aware service for mobile context-aware computing applications by the example of google android. In: 2012 IEEE International Conference on Pervasive Computing and Communications Workshops (PERCOM Workshops), pp. 76–81, March 2012

19. Löwe, R., Mandl, P., Weber, M.: Supporting generic context-aware applications for mobile devices. In: 2013 IEEE International Conference on Pervasive Computing and Communications Workshops (PERCOM Workshops), pp. 97–102, March 2013

20. Maingret, B., Mouël, F.L., Ponge, J., Stouls, N., Cao, J., Loiseau, Y.: Towards a decoupled context-oriented programming language for the internet of things. In: International Workshop on Context-Oriented Programming, pp. 1–6. ACM (2015)

21. Makris, P., Skoutas, D.N., Skianis, C.: A survey on context-aware mobile and wireless networking: on networking and computing environments' integration. IEEE Commun. Surv. Tutorials **15**(1), 362–386 (2013)

22. Marzano, S.: The New Everyday: Views on Ambient Intelligence. 010 Publishers, Rotterdam (2003)

23. Merlo, A., Migliardi, M., Caviglione, L.: A survey on energy-aware security mechanisms. Pervasive Mob. Comput. **24**, 77–90 (2015). http://www.sciencedirect.com/science/article/pii/S1574119215000929. Special Issue on Secure Ubiquitous Computing

24. Mikkonen, T., Systä, K., Pautasso, C.: Towards liquid web applications. In: Cimiano, P., Frasincar, F., Houben, G.-J., Schwabe, D. (eds.) ICWE 2015. LNCS, vol. 9114, pp. 134–143. Springer, Heidelberg (2015)

25. Miranda, J., Makitalo, N., Garcia-Alonso, J., Berrocal, J., Mikkonen, T., Canal, C., Murillo, J.: From the internet of things to the internet of people. Internet Comput. IEEE **19**(2), 40–47 (2015)

26. Olaru, A., Florea, A.M., Fallah Seghrouchni, A.: A context-aware multi-agent system as a middleware for ambient intelligence. Mob. Netw. Appl. **18**(3), 429–443 (2012). http://dx.doi.org/10.1007/s11036-012-0408-9

27. Park, H.-S., Oh, K., Cho, S.-B.: Bayesian network-based high-level context recognition for mobile context sharing in cyber-physical system. Int. J. Distrib. Sensor Netw. **7** (2011). doi:10.1155/2011/650387

28. Pitichat, T.: Smartphones in the workplace: changing organizational behavior, transforming the future. LUX: J. Transdisciplinary Writ. Res. Claremont Graduate Univ. **3**(1) (2013). http://scholarship.claremont.edu/lux/vol3/iss1/13

29. Qian, H., Andresen, D.: Extending mobile device's battery life by offloading computation to cloud. In: Abadi, A., Dig, D., Dubinsky, Y. (eds.) MOBILESoft 2015, pp. 150–151. Piscataway, IEEE (2015)

30. Ranganathan, A., Campbell, R.H.: A middleware for context-aware agents in ubiquitous computing environments. In: Endler, M., Schmidt, D.C. (eds.) Middleware

2003. LNCS, vol. 2672, pp. 143–161. Springer, Heidelberg (2003). doi:10.1007/3-540-44892-6_8

31. Roda, C., Rodríguez, A., López-Jaquero, V., González, P., Navarro, E.: A multi-agent system in ambient intelligence for the physical rehabilitation of older people. In: Bajo, J., Hernández, J.Z., Mathieu, P., Campbell, A., Fernández-Caballero, A., Moreno, M.N., Julián, V., Alonso Betanzos, A., Jiménez-López, M.D., Botti, V. (eds.) Trends in Practical Applications of Agents, Multi-agent Systems and Sustainability. AISC, vol. 372, pp. 113–124. Springer, Heidelberg (2015)

32. Sorici, A., Picard, G., Boissier, O., Florea, A.: Multi-agent based flexible deployment of context management in ambient intelligence applications. In: Demazeau, Y., Decker, K.S., Bajo Pérez, J., De la Prieta, F. (eds.) PAAMS 2015. LNCS, vol. 9086, pp. 225–239. Springer, Heidelberg (2015)

33. Taivalsaari, A., Mikkonen, T., Systä, K.: Liquid software manifesto: the era of multiple device ownership and its implications for software architecture In: IEEE 38th Annual Computer Software and Applications Conference, COMPSAC 2014, Vasteras, Sweden, 21–25 July 2014, pp. 338–343. IEEE (2014). http://dx.doi.org/10.1109/COMPSAC.2014.56

34. Weiser, M.: The computer for the 21st century. SIGMOBILE Mob. Comput. Commun. Rev. 3(3), 3–11 (1999). http://doi.acm.org/10.1145/329124.329126

35. Yeo, K.S., Chian, M.C., Ng, T.C.W., Tuan, D.A.: Internet of things: trends, challenges and applications. In: 2014 14th International Symposium on Integrated Circuits (ISIC), pp. 568–571 (2014)

Synchronizing Application State Using Virtual DOM Trees

Jari-Pekka Voutilainen[1]($\boxtimes$), Tommi Mikkonen[2], and Kari Systä[2]

[1] Gofore Ltd., Hämeenkatu 16, 33200 Tampere, Finland
jari.voutilainen@iki.fi
[2] Tampere University of Technology, Korkeakoulunkatu 1, 33720 Tampere, Finland
{tommi.mikkonen,kari.systa}@tut.fi

Abstract. We will all soon have numerous computing devices we use every day interchangeably. Liquid software, a concept where software is allowed to flow from one computer to another, is a programming framework that aims at simplifying the development and use of such multi-device software. The existing research has discovered three major architecture challenges for liquid software: (1) adaptation of the user interface to different devices, (2) availability of the relevant data in all devices, and (3) transfer of the application state. This paper addresses the last challenge and differs from the earlier work by concentrating in application state that is in the DOM tree, a key element in today's Web applications.

Keywords: Web programming · Multi-device ownership · Experience roaming · Liquid software

1 Introduction

Managing computers and software that runs in them is complicated when using multiple devices interchangeably. While multi-device programming has been proposed [1], we assume using PCs and mobile devices as the primary devices, with the need to maintain each of them separately and independently of each other [2]. This includes both data and applications that are run in the devices. While cloud computing helps with data, reinstalling applications, setting up accounts, and other complications will however plague our computing experience for years to come.

Liquid software, a concept where software is allowed to flow from one computer to another [3–5], is a programming framework that aims at simplifying the development of multi-device software. Instead of treating applications as device-specific nuisance that must be installed to all of them, the framework enables switching devices while still running the same application. Furthermore, we assume that it is thus achieved using only web technologies [6].

In this paper, we extend the technology basis [7] with a new DOM tree-based synchronization between different browsers. The paper is structured as

© Springer International Publishing AG 2016
S. Casteleyn et al. (Eds.): ICWE 2016 Workshops, LNCS 9881, pp. 142–154, 2016.
DOI: 10.1007/978-3-319-46963-8_12

follows. In Sect. 2, we discuss the earlier work and background of this paper. In Sect. 3 we define the goals of the framework. In Sect. 4 we introduce the new implementation, building on DOM trees. In Sect. 5 we look at our example application. In Sect. 6, we provide an extended discussion regarding the lessons learned regarding the design, and draw some final conclusions.

2   Background and Motivation

Experiences gained with numerous experiments, including for example Cloudberry [8], HTML5 Mobile Agents [9], and Cloud Browser [10], led us to consider a different type of computing paradigm, where applications are not artificially tied to a particular computer. Instead, they could flow from one computer to another, in a fashion that is casual and free from hassle from end-user perspective – or, as put in our earlier papers, in a liquid fashion [5]. Our earlier research has motivated us to build a application framework that would make development of liquid applications easier. The same experiments showed us that rich, powerful and especially portable applications can be implemented by using standard Web technologies like HTML, CSS and JavaScript.

The software framework for Liquid applications includes three main subsystems [6]:

- Adaptation of the user interface to different devices and contexts. While the functions of an application may remain the same, the devices used to run the application may differ in terms of display and input technologies. In addition, the users' ability and interest to pay constant attention differ according to the context.
- Synchronization of the users' persistent data and content. The applications need to have an access to files like pictures, documents and PIM (Personal Information Management) data. This content is commonly stored in cloud-based services instead of the local storage of the devices. Unfortunately, these cloud-based storage systems are generally limited to single applications, Internet services, or to ecosystems built on certain operating systems. Some research like data API of Cloudberry mentioned above and EDB [11] address these problems by providing programming APIs and efficient synchronization mechanisms. Similar functionality could also be achieved by storing data in BaaS (Backend-as-a-Service) systems like Firebase[1] but those systems are not optimized for files.
- Synchronization of the state of the applications that move during execution or for applications that run on multiple devices at the same time. This state information includes values of relevant variables and control state. In case of Web applications the state could include JavaScript variables and content in the DOM tree. Since the execution model of Web applications is based on reasonable short events handlers, and especially if we assume that Web Workers are not used and applications should not move during incomplete I/O operation, no control state has to be included.

---

[1] https://www.firebase.com/.

In this paper we concentrate on the last item, i.e., synchronization of the application state and place the focus on technology that is needed to implement state synchronization. Other topics have been addressed in our earlier work [2,5,6,12]. While designing a architecture for state synchronization, the level of synchronization needs to be decided. These levels would range from lightweight variable transfer to complete application transfer. The most complete mechanism of migrating application state from one computer to another would be to transfer the complete memory space of the application. However, this is not technically feasible for a number of reasons:

- The amount of data to be moved is too big for almost any non-trivial application.
- Internal data structures, stored in the memory, cannot be moved as such, since they depend on underlying platform, as well as browser and virtual machine implementation.
- Present browser implementations do not support direct, serialized memory dumps.
- The application specific needs are not taken into account. In reality the set of synchronized data depends on the application.
- The content of the memory depends also on the browser implementation, and complex conversion mechanisms would be needed.

Over time, we have been working on several approaches for synchronization of the application state. All these approaches have their strengths and weaknesses. The approaches and their characteristics will be addressed next.

In Lively3D [13] we designed a 3D environment within browser window, where applications were rendered to 3D objects. The framework included utilities for serialization and deserialization of Javascipt objects that contained parts of application state. The serialized object was migrated to another browser through Dropbox or MongoDB database. In this framework the developer had to define the structure and content of JavaScript state objects as well as the concrete serialization and deserialization functions.

In Cloudberry [8], Cloudbrowser [10], and in some student projects, we have used BaaS based (Backend-as-a-Service) synchronization – either proprietary solutions built for the exact purpose, like in Cloudberry, or publicly available BaaS systems. Although these experiments have shown that the applications can be made movable with these solutions, the approach assumes that the applications uses the BaaS system explicitly store the critical state information.

In HTML5 Mobile agents [9] the applications mark the JavaScript variables to be synchronized. The values of the marked variables are then automatically serialized and migrated as part of the moving agent. This approach works well for cases where applications are not run simultaneously and the DOM tree is not used for storing relevant state information. Unfortunately, many programming paradigms and toolkits assume that part of the state is stored in DOM. These limitations lead us to think about a solution that can provide more automatic synchronization of data and also include data in the DOM tree.

Another design decision is the communication topology. In many of our experiments we have used a centralized topology. In that approach all state updates are stored in a centralized server from which other instances receive the state updates. In cases were applications are run simultaneously on multiple devices some peer-to-peer mechanisms could be used for synchronization, too. In our initial experiments we have used WebRTC [14] to implement a peer-to-peer synchronization routine. This approach is not applicable to cases where applications are not running on multiple devices simultaneously. Furthermore, such approach faces some implementation problems due to network infrastructures; for instance, NAT (Network Address Translation) often make use of the peer-to-peer approach difficult.

As listed above there are several approaches to implement roaming of the applications from device to another. The options differ from each other in several ways:

- support for simultaneous, sequential or both scenarios;
- need to a have centralized server;
- affect to programming model, where the state can be, selection of UI libraries, and other constraints.

3   Goals and the Core Idea

This work is a part of larger research aiming at a complete application framework for liquid applications where the goal is to minimize the additional work by the developer. Thus, the requirements for the framework are:

- Development of applications for the framework should be simple. The devel oper should not need to design the application to be Liquid and most of the existing applications should be able to use the framework in some extent.
- The framework should be based on existing browser implementations. The user should not need to install a custom browser or fiddle with the settings. Any modern browser should be usable as long as it supports the Web Standards used in the framework.
- The applications using the framework should be device agnostic. Part of the Liquid Software Manifesto is multi-device ownership, Liquid applications should be usable in desktops, mobile phones and tablets. Therefore the applications using the framework must be usable also in mobile devices and the migration of applications state should not differentiate between these.

The application state of any modern client-side web application is stored in both JavaScript namespace and the DOM tree of the browser window. This may be result of explicit choice by the developers or consequence of the libraries used in the implementation. We see that a system that can serialize content from both JavaScript variables and DOM tree provides the most flexible solution for liquid web applications since it supports wide range of different frameworks and programming styles. In the work reported in this paper we complement our earlier work on serializing state in the JavaScript variables with techniques to migrate parts of the DOM tree from one browser window or device to another and restoring it to the same state.

Although the browsers provide JavaScript APIs to read the content of the DOM tree, copying of the whole tree after each state change is not feasible. The DOM tree of the application is often large and most of it remains unchanged during execution of application. In most applications some content of the DOM tree also reside in JavaScript variables, which could be used to restore the DOM tree at least in some extent. Without designing the application to store all the DOM content in JavaScript variables, relying only to the JavaScript variables is not feasible to transfer DOM tree states.

This paper proposes a new mechanism to serialize data information from the DOM tree by using a concept called *Virtual DOM trees*. Virtual DOM is a novel technology popularized by React.js[2]. The basic idea of Virtual DOM is that rather than touching the DOM tree directly, application builds an abstract version of the tree. This abstraction makes DOM manipulation significantly faster when is coupled with efficient comparison algorithms and operation on selective sub-trees. Other implementations of Virtual DOM include virtual-dom[3], Mithril[4] and Bobril[5].

Virtual DOM enables manipulation of the DOM trees in JavaScript namespace. We can create virtual DOM trees in JavaScript variables, modify them, compare two different virtual trees and apply these comparisons back to the actual DOM tree. Similar technology under standardization is Shadow DOM [15], which encapsulates multiple DOM trees within one document. However, this technology does not provide means to compare the current DOM tree to another tree and therefore it cannot be used in our approach as such.

4   Design and Implementation

We present our proof-of-concept implementation of the Liquid Application Framework, later named Liquid.js. Liquid.js provides APIs for application developer to use, to enable DOM tree migrations in easy to use manner. It should be noted that our ultimate goal is a complete Liquid.js that supports several aspects of making application liquid, but the proof-of-concept discussed in this paper concentrates on the reported mechanism for state synchronization.

The cornerstone of our approach is the initial state of the DOM trees. When an application is loaded and the browser throws a *load* event, the initial state is formed as a Virtual DOM tree and stored within the framework for further use. Later, during migration, we only transfer differences between the original Virtual DOM tree and the current DOM to minimize the amount of transferred data. When a migration of a state is initiated either by a user action or by some

---

[2] https://facebook.github.io/react/docs/glossary.html.
[3] https://github.com/Matt-Esch/virtual-dom.
[4] http://mithril.js.org/.
[5] https://github.com/Bobris/Bobril.

external event, the current state of the DOM tree is compared against the stored initial state using comparison algorithm provided by the *virtual-dom* -library. This comparison produces a set of patch operations that encapsulates the needed modifications from the initial DOM tree to the current tree. To minimize the data needed for migration, the prototype objects (named *VirtualNode*, *VirtualPatch* and *VirtualText*) are removed before sending and re-created after migration in the new browser.

After the prototypes have been recreated, the received patch operations are applied to the initial state virtual DOM tree. This tree is then applied to the actual DOM tree. This results in exactly the same state as in the origin browser. This process is also presented in Fig. 1.

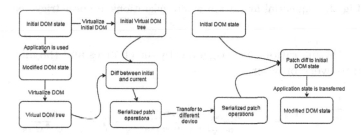

Fig. 1. Serialization of DOM tree to different browsers.

Since the patch operations assume the commonly known initial state of the DOM tree, the destination browser needs to be moved back to that initial state before applying the patches. After the destination browser is in known initial state, the deserialized migration object with the recreated prototype objects is applied to the DOM tree which results to the synchronized DOM state between the browsers.

There are several options for the architecture and implementation of the network. In our proof-of-concept the synchronization is based on minimal store-and-forward server and WebSockets for inter-browser communication. The centralized server is the bare minimum needed for the migration to work. It provides functions to query the browsers that have initialized Liquid.js applications and to transfer migration objects between browsers. The migration objects can be sent to single browser or all of them.

Our demo application, presented in detail in Sect. 5, is loaded into multiple browser windows in different devices. In each browser the initial state is stored during the initialization of Liquid.js. In our demo, the migrations are triggered by user action, but since Liquid.js provides JavaScript function for triggering the migration, almost any browser event could be used for it.

Due to the discarding of the current state during migration in receiving browser, Liquid.js essentially implements the sequential paradigm of Liquid Software. But because the application is not removed from the source browser, simultaneous paradigm could be implemented with migrating every time the state changes in one of the browsers. For fully implementing simultaneous screening, merging of the states should be implemented instead of reverting to the known initial state. Examples of both scenarios are provided in Figs. 2 and 3, where state is migrated between three different browsers. In Fig. 2, the migrations are user triggered, and there, the user chooses the destination browser.

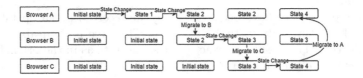

Fig. 2. Sequential migration where migrations are user triggered.

In Fig. 3, the migrations are triggered by the changes and the migrations are transferred to every browser.

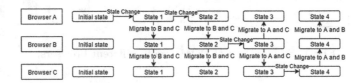

Fig. 3. Simultaneous migration where migrations are triggered by state changes.

The migration object formed from the Virtual DOM tree contains information only found from the real DOM tree, for example class names, identifiers and identities, values, and other attributes. Anything related to the DOM tree, but actually residing in the JavaScript namespace will not get serialized with the DOM tree. Most common case is event listeners that are bound to the DOM elements. After migration the DOM tree would be the same, but the functionality bound to the DOM elements would have disappeared. An example of the DOM tree has been provided in Fig. 4. Everything not present in the example DOM elements, are not migrated with virtual DOM.

Since only the data residing in the DOM tree is migrated through virtual DOM trees, Liquid.js has functionality to register event listeners that should be serialized with the DOM tree and rebound during creation of DOM elements. This was implemented so that application does not need to pollute the global scope with function names that could be visible in a much smaller scope. Example of application that uses Liquid.js to enable migration of DOM trees is provided in Listing 1. This listing presents only the lines of code needed to enable successful migration and it uses the default implementations to simplify the example.

```
▶ <div class="row">…</div>
  ▶ <div class="row">…</div>
    <hr>
  ▼ <ul class="list-group todo-list">
    ▼ <li class="list-group-item">
      ▶ <input type="checkbox" class="toggle" data-handler="toggleClick">…</input>
        <label class="done">item 1</label>
        <button class="destroy" data-handler="destroyClick"></button>
      </li>
    ▶ <li class="list-group-item">…</li>
    ▶ <li class="list-group-item">…</li>
    </ul>
    ::after
  </div>
  <script src="https://ajax.googleapis.com/ajax/libs/jquery/1.11.3/jquery.min.js"></script>
  <script src="bootstrap/dist/js/bootstrap.min.js"></script>
```

Fig. 4. A part of the DOM tree as seen in chrome dev tools.

```
var Liquid = require('liquid.js');
var liquid = new Liquid();

var checkbox = document.createElement('input');
checkbox.addEventListener('click', someFunction);
checkbox.setAttribute('data-handler', 'toggleClick');

liquid.registerHandler('toggleClick', {'click': someFunction});
```

Listing 1: Example application using Liquid.js

Liquid.js consists of multiple modules designed to handle different aspects of the framework. These modules interact with each other and third party developed open source modules, which each have their own dependencies. In Fig. 5 a graph of the modules is presented with explanations if the module is implemented with the Liquid.js framework or if it is developed by third party. Each box in the figure is a single module and the arrows describe dependencies between them. Third party modules are available in npm[6] and Github[7].

The description of modules is given in following:

- *Liquid.js* is the the main module, which encapsulates the functionality of other modules. This also provides interfaces for the application developer.
- *Liquid.js / UI* provides optional User Interface elements, which can be used to control migrations.
- *Liquid.js / Variables* contain functions for variable and function registrations. These are required to enable rebounding of functions after migration.
- *Liquid.js / VDOM* contains Virtual DOM tree related functions.
- *vdom-serialize* provides functionality for serialization and deserialization of Virtual DOM trees.
- *vdom-virtualize* enables Virtual DOM tree creation from DOM tree.
- *virtual-dom* handles Virtual DOM related algorithms: comparison and merging Virtual DOM to DOM tree.

---

[6] https://www.npmjs.com/.
[7] https://github.com/.

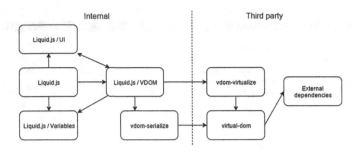

Fig. 5. Architecture of Liquid.js. Each box is a module available in npm or Github.

Even though Liquid.js is built using WebSockets, practically any communication mechanisms could be used as the communication channel for the state migration. The constructed data structure is a standard JSON object, which is serialized by Liquid.js, and it could be transferred in a number of ways, including server-side databases, BaaS solutions, and even WebRTC. The framework is designed to support other transfer methods and the application developer can choose which one to use. By choosing different transfer methods, application developer can implement more or less real-time synchronizations for different needs of applications.

The application needs to conform to the limitations of DOM trees and APIs that Liquid.js provides for enabling state migration. If the following requirements are fulfilled, the migration will be successful without any complications:

- The DOM tree is built so that its state is applicable for migration as it is at any given time.
- Event handler functions and variables that are not in global scope, are required to be registered to Liquid.js.
- The receiving browser's Liquid.js reverts application to initial state before applying the migrating application state to avoid merging complications which are out of scope of Liquid.js.

Responsive user interfaces are out of scope of Liquid.js, but they can be used in applications if the requirements listed above are met. For example Bootstrap[8] and Foundation[9] both implement responsive UIs using CSS classes, which are stored within the DOM tree, so these would work without a problem.

5   Example Application

Our example application is a basic todo list application. It has functionality for adding items, marking them done and removing them completely. All the application state is stored within the DOM tree for the sake of the demo, the JavaScript namespace does not contain anything that is not present right

---

[8] http://getbootstrap.com/.
[9] http://foundation.zurb.com/.

after application initialization. Whole application is built using Browserify[10] which enables using require function in frontend code. A screenshot and application structure is presented in Fig. 6.

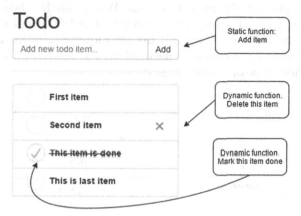

Fig. 6. Demo application and its functions without Liquid.js.

The liquid framework need to be initializes before use as presented in Listing 2. By default this initializes WebSockets for communication and adds user interface elements for management of the migrations. During initialization the initial state is also stored for future use, so Liquid.js should be initialized in *load*-event which is thrown by the browser when the application has finished loading. This enables virtual DOM migrations between browsers.

```
var Liquid = require('liquid.js');
var liquid = new Liquid();
```

Listing 2: Initialization of Liquid.js

As stated above while migrating only the virtual DOM, the bindings to dynamic JavaScript event handlers are lost since they are not residing in the DOM. These need to be registered to Liquid.js, so that they are serialized with the virtual DOM. In registration Liquid.js needs to know in what event it should bind the function after migration. Registered event handlers are given a name in the DOM, this is saved to the *data-handler* attribute of the DOM element. Example of these is provided in Listing 3.

```
// Add data-handler="removeItem" to removeButton
removeButton.setAttribute('data-handler', 'removeItem');

// Register removeTodo event handler to bind
//click-event to DOM element with data-handler removeItem
liquid.registerHandler('removeItem', {'click': removeTodo});
```

Listing 3: Registration of event handler for serialization.

---

[10] http://browserify.org/.

While migrating Liquid.js removes its own user interface elements, since they are not part of the application and we do not want to migrate those. After removing, Liquid.js serializes the differences in DOM compared to the initial state and the registered dynamic functions. When this has been migrated to another browser, the differences are applied to the DOM and event handlers are rebound to the DOM elements. Finally the Liquid.js user interface elements are recreated. Figure 7 presents screenshot of the application with Liquid.js user interface and their removal during migration.

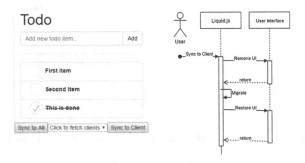

Fig. 7. User interface elements of Liquid.js and their removal during migration.

6   Conclusions

The emerging paradigm of programmable world, where all our objects act as one [16], means that all the devices can no longer be treated as individual computers – all their data, content, and programs simply cannot be installed separately in each device by the end user. Instead, we need a new type of programming paradigm we call liquid software, where applications from one device can migrate to another one, and continue their execution in the new context.

In this paper, we propose a framework for liquid web applications. The framework is based on synchronizing DOM trees to migrate the state of a web application, and it contrasts our earlier experiments [2,5,6,12] by proposing a direction that is closer in spirit to web designs than programming approaches. While the presented implementation is based on WebSockets, this is not an essential design restriction, as (almost) any other protocol could be used as the transport mechanism.

Liquid.js implements proof-of-concept prototype of one part of the liquid software vision. It demonstrates that storing vital information of application to the DOM tree and migrating the tree to another browser is possible and without the need of rebuilding the application by the developer. In numerous practical cases, applications would not solely rely on either DOM trees nor JavaScript namespace to store the state of the application. In practice applications would store some of the state to the DOM tree and some to JavaScript namespace.

During the implementation of Liquid.js, it was found out that in the current web programming model, the DOM tree, and JavaScript namespace are deeply mingled. Since virtual DOM only interacts with the DOM tree, all the information stored in JavaScript namespace that were not there during the initialization of application, were lost during migration in our early experiments. We needed a functionality to specify important variables and functions to be serialized with the DOM tree, so that after migration the application would be practically the same with state and functionality.

Our design implements migration of applications with minimal overhead to application code. With only few functions to call when registering relevant variables and functions related to the DOM tree, it is simple to enable migration while developing applications. Even though there probably still are some issues to be solved in our proof of concept, Liquid.js presents DOM trees as viable option to store application state in Liquid Software. It has been common practice as long as there have been web applications and it should not be a problem in Liquid Software either. All code for our proof-of-concept is available at https://github.com/Zharktas/liquid.js and for the demo application at https://github.com/Zharktas/Liquid-Todo.

In further research we would develop more complete solution where we would implement results from our previous findings to a single application framework. This would include DOM tree migrations, mechanisms for JavaScript namespace transfer and merging of application states in simultaneous screening instead of replacement.

Related studies have been made: in [17] Bellucci et al. researched serialization of JavaScript namespace and implemented serialization functions for multiple standard JavaScript objects like timers and dates. In [18] Ghiani et al. proposed a proxy server implementation which was used to transfer browser session from user to user. Finally in [19] Klokmose et al. implemented framework called Webstrates which executed web applications in iframe elements and the parent frame handled state synchronization between browsers.

Acknowledgments. This work has been partially supported by Academy of Finland (projects 283276 and 295913).

References

1. Mäkitalo, N., Pääkkö, J., Raatikainen, M., Myllärniemi, V., Aaltonen, T., Leppänen, T., Männistö, T., Mikkonen, T.: Social devices: collaborative co-located interactions in a mobile cloud. In: Proceedings of the 11th International Conference on Mobile and Ubiquitous Multimedia, p. 10. ACM (2012)
2. Taivalsaari, A., Mikkonen, T.: From apps to liquid multi-device software. Proc. Comput. Sci. **56**, 34–40 (2015)
3. Hartman, J., Manber, U., Peterson, L., Proebsting, T.: Liquid software: a new paradigm for networked systems. Technical report, Technical report 96 (1996)
4. Hartman, J.J., Bigot, P., Bridges, P., Montz, B., Piltz, R., Spatscheck, O., Proebsting, T., Peterson, L.L., Bavier, A., et al.: Joust: a platform for liquid software. Computer **32**(4), 50–56 (1999)

5. Taivalsaari, A., Mikkonen, T., Systa, K.: Liquid software manifesto: the era of multiple device ownership and its implications for software architecture. In: 2014 IEEE 38th Annual Computer Software and Applications Conference (COMPSAC), pp. 338–343. IEEE (2014)
6. Mikkonen, T., Systä, K., Pautasso, C.: Towards liquid web applications. In: Cimiano, P., Frasincar, F., Houben, G.-J., Schwabe, D. (eds.) ICWE 2015. LNCS, vol. 9114, pp. 134–143. Springer, Heidelberg (2015)
7. Gallidabino, A., Pautasso, C., Ilvonen, V., Mikkonen, T., Systä, K., Voutilainen, J.P., Taivalsaari, A.: On the architecture of liquid software: technology alternatives and design space
8. Taivalsaari, A., Systä, K.: Cloudberry: an HTML5 cloud phone platform for mobile devices. IEEE Softw. **29**(4), 40–45 (2012)
9. Systä, K., Mikkonen, T., Järvenpää, L.: HTML5 agents: mobile agents for the web. In: Krempels, K.-H., Stocker, A. (eds.) WEBIST 2013. LNBIP, vol. 189, pp. 53–67. Springer, Heidelberg (2014)
10. Taivalsaari, A., Mikkonen, T., Systä, K.: Cloud browser: enhancing the web browser with cloud sessions and downloadable user interface. In: Park, J.J.J.H., Arabnia, H.R., Kim, C., Shi, W., Gil, J.-M. (eds.) GPC 2013. LNCS, vol. 7861, pp. 224–233. Springer, Heidelberg (2013)
11. Koskimies, O., Wikman, J., Mikola, T., Taivalsaari, A.: EDB: a multi-master database for liquid multi-device software. In: Proceedings of the Second ACM International Conference on Mobile Software Engineering and Systems. MOBILESoft 2015, pp. 125–128. IEEE Press, Piscataway (2015)
12. Voutilainen, J.P., Salonen, J., Mikkonen, T.: On the design of a responsive user interface for a multi-device web service. In: Proceedings of the Second ACM International Conference on Mobile Software Engineering and Systems, pp. 60–63. IEEE Press (2015)
13. Voutilainen, J.P., Mattila, A.L., Mikkonen, T.: Lively3D: building a 3D desktop environment as a single page application. Acta Cybernetica **21**, 291–306 (2014)
14. W3C WebRTC Working Group: WebRTC 1.0: real-time communication between browsers (2016)
15. W3C Web Platform Working Group: Shadow DOM (2015)
16. Wasik, B.: In the programmable world, all our objects will act as one. Wired (last modified 14 May 2013), p. 462 (2013)
17. Bellucci, F., Ghiani, G., Paternò, F., Santoro, C.: Engineering javascript state persistence of web applications migrating across multiple devices. In: Proceedings of the 3rd ACM SIGCHI Symposium on Engineering Interactive Computing Systems, pp. 105–110. ACM (2011)
18. Ghiani, G., Paternò, F., Santoro, C.: Push and pull of web user interfaces in multi-device environments. In: Proceedings of the International Working Conference on Advanced Visual Interfaces, pp. 10–17. ACM (2012)
19. Klokmose, C.N., Eagan, J.R., Baader, S., Mackay, W., Beaudouin-Lafon, M.: Webstrates: shareable dynamic media. In: Proceedings of the 28th Annual ACM Symposium on User Interface Software & #38; Technology. UIST 2015, pp. 280–290. ACM, New York (2015)

5th Workshop on Distributed User Interfaces: Distributing Interaction (DUI 2016)

AttributeLinking: Exploiting Attributes for Inter-component Communication

Michael Krug[(✉)] and Martin Gaedke

Technische Universität Chemnitz, Chemnitz, Germany
{michael.krug,martin.gaedke}@informatik.tu-chemnitz.de

Abstract. In this paper, we propose exploiting attributes of client-side web components to provide inter-component communication by external configuration. With the standardization of WebComponents, the Web is finally getting a uniform way to define and use client-side components. We determined that DOM elements already provide a standard configuration interface: attributes. Using the WebComponents technologies for state-of-the-art user-interface components, attributes can also act as output interfaces. By providing an Attribute-Link component, new web applications can be composed directly in the markup without knowledge of JavaScript. With the integration of a multi-device supporting Messaging-Service, components can be even linked across multiple connected devices. This enables the development of distributed user interfaces.

Keywords: Web components · Web application development · Composition · Distributed user interfaces · Reusable components

1 Introduction

Component-based application development is based on the composition of multiple components that are reused and combined to form new applications. Applying this to the Web creates new challenges, like the definition of independent components and loosely coupling. A lot of approaches were developed in the last years, but a standardized way of defining and using components for client-side Web application development was missing. Recently, the W3C is working on creating a new set of standards for components for the Web called *WebComponents*[1]. With those standards developers are now able to define new types of DOM elements with custom functionality. Similar to UI mashups, where user-interface components were mostly referred to as widgets, inter-component communication, which is required for application development by composition, is a central problem. This is unfortunately not covered by the W3C specifications. We analyzed different DOM elements and determined that most of them provide an accessible (input) interface, which is mainly used for configuration: *attributes*. Consequently, this interface is also available for WebComponents.

---

[1] WebComponents - W3C Wiki https://www.w3.org/wiki/WebComponents/.

© Springer International Publishing AG 2016
S. Casteleyn et al. (Eds.): ICWE 2016 Workshops, LNCS 9881, pp. 157–161, 2016.
DOI: 10.1007/978-3-319-46963-8_13

```
<geo-coder  address="Chemnitz"// input interface
            lat="50.83"         // output interface
            lng="12.92">        // output interface
</geo-coder>
```

Listing 1. Geo-Coder component with multiple attributes

With an accordingly configuration, attributes of WebComponents can also be used as output interfaces. To explain this idea, an example in Listing 1 is stated, where a *Geo-Coder* component is shown that converts an address into geographic coordinates (latitude, longitude), e.g., to show a place on a map. Thus, the attribute *address* is used as an input interface and *lat* and *lng* are the resulting output variables. If we now assume that there is also a *Map* component with latitude and longitude as input variables mapped to the according attributes, a composition of those elements can be achieved by connecting the output variables of the *Geo-Coder* to the input ones of the *Map*.

Therefore, we want to encourage developers to use attributes as communication interfaces when creating new WebComponents. By enabling external configuration of inter-component data exchange the components do not need to include any communication functionality by themselves. Thereby, components stay reusable and do not need knowledge of other components.

2   Related Work

Polymer, a WebComponent framework from Google, supports data bindings within markup through placeholders in attributes and by placing target and source within a binding element. There are also several JavaScript frameworks that support data binding, e.g., AngularJS, Knockout or Rivets.js. They need an attached data model to specifically bind data to the element's content or attributes. Furthermore, some publish/subscribe or event-based communication approaches for components or widgets, like presented in [1,2] and [3], exist. They have the disadvantage of configuring communication aspects within the components. There are only limited options to configure which components shall exchange data. Data transformation and message distribution for multi-device applications is not supported.

3   The AttributeLinking Approach

Based on the assumption that WebComponents are DOM-Elements that provide their in- and output interfaces through attributes, we enable inter-component communication by linking attributes through external configuration to facilitate web application development by composition. How to connect those interfaces?

3.1   The Attribute-Link Component

We propose to connect the attribute interfaces of components using a stand-alone *Attribute-Link* component, which is configured using three attributes: *source*, *target* and an optional *transformation* (see Listing 2). The *Attribute-Link* component is also implemented as a WebComponent. Thus, it can be natively used in web browsers supporting those standards and can be deployed directly in the markup without knowledge of JavaScript. To address an element we propose to use the established *CSS selector* syntax[2]. This syntax is applied to both the source as well as the target selector. Since CSS selectors can return multiple elements, we also support multiple elements as target and source. Enabling a cardinality of both ends from 1 to n. The attribute is addressed by its name separated by an @ sign. Example: `geo-coder#id1@address`. The *Attribute-Link* can be included multiple times in the web application to define a complex inter-component communication setup. To watch for attribute changes without affecting the responsiveness of the application we use native DOM mutation observers[3] that set up microtasks and provide a callback for registered changes of the specific element's attribute. If a transformation function was specified, it will be applied and the resulting value is propagated to the defined attribute of the according target element.

```
<attribute-link  source="selector@attribute"
                 target="selector@attribute">
                 transformation="JavaScript-Code"/*Optional*/>
</attribute-link>
```

Listing 2. Syntax of the proposed Attribute-Link component

3.2   Distributing Attribute Changes

In our approach, we also focus on the development distributed user interfaces. Thus, we propose to additionally address target components on connected devices. This is achieved using a *Messaging-Service* component and a *Messaging-Server* we provide. Our presented *Attribute-Link* component seamlessly integrates with the *Messaging-Service* by placing it in the DOM as a child element like display in Listing 3. It notifies the *Messaging-Service* of changed attributes that shall be distributed by dispatching a custom event. This event contains the target element selector, the attribute name and the new value to be set. Using the *WebSocket protocol*[4] the *Messaging-Server* is contacted to distribute this message to other connected devices (browsers running the application with the same endpoint configured that share the same context) (cf Fig. 1). The other devices also need to have instantiated an accordingly configured *Messaging-Service* component that will then propagate the received messages to the target elements in their DOM. We also provide a configuration option to only target remote components.

---

[2] Selectors Level 3 https://www.w3.org/TR/selectors/.
[3] DOM Standard https://dom.spec.whatwg.org/#mutation-observers.
[4] The WebSocket Protocol http://tools.ietf.org/html/rfc6455.

```
<messaging-service endpoint="protocol://address:port"
                   session="Session-Identifier">
    <attribute-link [...]></attribute-link>
</messaging-service>
```

Listing 3. Syntax of the Messaging-Service in combination with an Attribute-Link

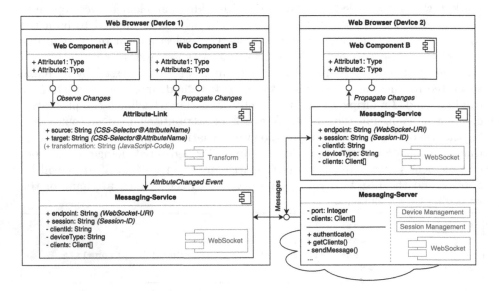

Fig. 1. Components of the AttributeLinking approach

4   Conclusion

In this paper, we proposed the external linking of attributes of DOM-based components to enable the composition of new web applications. Consequently, we want to start a discussion of the usage of attributes as in- and output interfaces for components in the Web. To demonstrate our idea, we presented an Attribute-Link component that is capable of observing and propagating attribute changes from and to multiple DOM elements. With an optional Messaging-Service those changes can also be distributed to connected devices. The concept of external configuration eliminates the need of modifying the components to enable communication. Further research will address how to enhance the support for distributed interfaces by providing more configuration options to e.g., select components on specific devices.

Online Demonstration: http://myvsr.eu/demo/dui/

References

1. Chudnovskyy, O., Müller, S., Gaedke, M.: Extending web standards-based widgets towards inter-widget communication. In: Grossniklaus, M., Wimmer, M. (eds.) ICWE Workshops 2012. LNCS, vol. 7703, pp. 93–96. Springer, Heidelberg (2012)
2. Krug, M., Gaedke, M.: SmartComposition: bringing component-based software engineering to the web. In: Proceedings of the 17th International Conference on Information Integration and Web-Based Applications and Services, pp. 474–477. ACM (2015)
3. Sire, S., Paquier, M., Vagner, A., Bogaerts, J.: A messaging API for inter-widgets communication. In: Proceedings of the 18th International Conference on World Wide Web, pp. 1115–1116. ACM (2009)

Improving Context-Awareness in Healthcare Through Distributed Interactions

Juan E. Garrido[1,2(✉)], Víctor M.R. Penichet[1,2], and María D. Lozano[1,2]

[1] Computer Science Research Institute, University of Castilla-La Mancha, Albacete, Spain
juanenrique.garrido@uclm.es
[2] Computer Systems Department, University of Castilla-La Mancha, Albacete, Spain
{victor.penichet,maria.lozano}@uclm.es

Abstract. Context-aware systems represent an appropriate tool for healthcare environments. The capability to offer necessary information and functionality on the basis of a user's context makes it possible to create better working environments for medical staff. Current technology supports greater customization, thus enabling the improvement of user interaction. This paper describes a significant step forward in the concept of context-awareness with a comprehensive solution: Ubi4Health. The solution enhances context-awareness by adapting the user experience in accordance with the device, interface and interaction mechanism used in a given context.

Keywords: Healthcare · Ubiquity · Context-awareness · DUI · HCI

1 Introduction

Medical staff have to manage complex situations [1] which involve: working with professionals from different fields; reaching agreement; performing multiple tasks; dealing with frequent contingencies that requires a constant adjustment of their actions because of the dynamic nature of their work [2]; and keeping in mind several pending tasks [3].

These conditions can be improved through context-aware [4] and ubiquitous [5] systems. Users can work without thinking about the system which is responsible for adapting itself to existing conditions at any time and under any circumstance. Generally speaking, context-aware systems adapt themselves to offer functionality and information that users need based on the context. However, current technological possibilities provide a chance of taking a step forward. The context offers valuable information, such as the physical capabilities of the user, which can be used to determine the appropriate interaction mechanism. In this sense, the distribution of multiple interaction methods across the working environment represents a good opportunity for context-awareness improvement.

The distribution of interaction mechanisms is directly related to multi-device environments. The possibility to interact with multiple devices offers more ways of interaction. But it also helps to continue improving context-aware capabilities. Users can interact with specific devices according to their needs, such as mobility conditions in

© Springer International Publishing AG 2016
S. Casteleyn et al. (Eds.): ICWE 2016 Workshops, LNCS 9881, pp. 162–166, 2016.
DOI: 10.1007/978-3-319-46963-8_14

healthcare [6]. At the same time, a multi-device environment adds a new possibility of improvement for context-aware systems: the use of the distribution of user interfaces. As users are provided with a wide range of devices, the system can adapt the interface to the hardware and also to the users' context.

Taking all the above into account, the authors have developed a ubiquitous and context-aware comprehensive solution for deployment in healthcare centres, named Ubi4Health [7]. The solution supports many functions, from task management, an alert notification mechanism, and falls and fainting detection to a rehabilitation assistant. This paper describes how Ubi4Health generates a novel context-aware environment for users (employees, patients and residents). The main contribution of this work is in the use of distributed interaction mechanisms, a multi-device environment and appropriate user interfaces to adapt the system to a user's context. These elements provide users with an experience that is adapted to their needs at the information and functionality level, and also to the way they interact with the system. The related works studied [6, 8–10] offer solutions to specific fields, as Ubi4Health does within its modules. However, none of them considers context-awareness as more than adapting data and functionality capabilities.

The following section summarizes the way Ubi4Health improves healthcare conditions thanks to its support for a novel development of context-awareness. Finally, we present our conclusions.

2 Ubi4Health: Improving and Making Progress with Traditional Context-Aware Systems

Ubi4Health is a novel, context-aware and ubiquitous comprehensive solution to enhance working conditions in healthcare environments. The solution has been developed modularly, and each module is related to a specific domain (Fig. 1): (1) *tasks module* to manage tasks (alerts are tasks with a high priority); (2) *rehabilitation module* to improve specific rehabilitation processes; and (3) *KFF (Kinect for Falls and Fainting) system* to automatically detect anomalous situations in the environment.

The context-awareness in Ubi4Health represents a significant contribution. The system offers functionality and information based on a user's context. However, the objective has been to go a step further. To that end, three features and capabilities are used together: *distributed multimodality, multi-device* and the *distribution of user interfaces*.

Firstly, Ubi4Health has a complete set of interaction mechanisms distributed over the healthcare environment (Fig. 1). On the one hand, the solution allows users to interact with traditional mechanisms (keyboard, mouse and touchscreen). These offer the possibility to work with tasks and rehabilitation modules regardless of the circumstances. However, there are situations in which it is necessary to make use of other interaction methods to enhance the user experience. In the tasks module there is a role named operator. This role has to control the completion of tasks, adequate staff distribution and the assignment of alerts. As a result, the amount of information to be managed is huge. Therefore, Ubi4Health incorporates the use of users' gaze to interact with the operator.

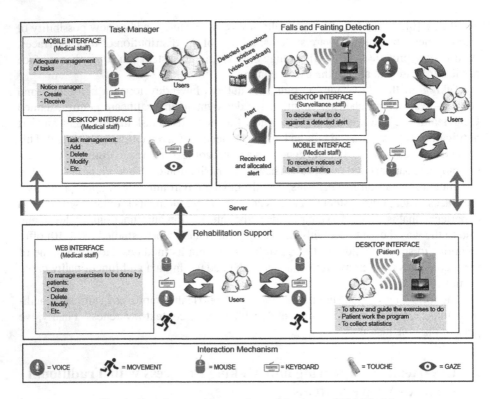

Fig. 1. Interfaces and interaction mechanisms of Ubi4Health

In this way, the system is able to know where the user is looking in order to identify loss of information on unattended displays. The system adapts its behaviour and aspect to a user's context as it generates notifications to assist the perception of information [11].

Movement-based interaction is an essential element. In the rehabilitation module, this interaction allows patients to complete their therapies at home. The system analyses a patient's movements and automatically guides and monitors. Also, this mechanism offers medical staff the possibility to define exercises to be performed by patients. In particular, medical staff must generate therapies, defined as posture repetitions. The postures can be created through a 3D designer, and also by the movement of the medical staff. Furthermore, in the KFF System, the solution uses movement interaction to continuously analyse a patient's/resident's postures. The objective is to detect falls and fainting, which is an important part of patient care.

The distributed multimodality offered by Ubi4Health is completed with the user's voice. The solution provides the possibility of control via voice for some interfaces. This feature helps in those situations in which the use of hands or the whole body is a complicated task. An example arises during the rehabilitation process. If a patient is performing an exercise, voice is a good way of controlling the system.

This multimodality is provided together with an important infrastructure which generates the second feature related to context-awareness improvement: the use of a wide device set. Ubi4Health offers pcs, laptops, tablets, PDAs, mobile phones, as well as Kinect cameras, web cameras and touchscreens. The result is the capacity to adapt the use of the solution to the appropriate device based on the user's needs. For example, the task module considers two user types with the role of employee who present different mobility needs. Firstly, there are users who are constantly in movement, such as doctors. They have to work with convenient devices to be carried with them and which can be used anywhere and at any time. Ubi4Health offers them mobile phones, tablets or PDAs. The other type of user has no movement needs, and is named operator. They are offered a pc, but with one peculiarity, namely a multi-display setting. The main reason for this comes from the fact of working with a large amount of information from different domains.

The last feature of Ubi4Health that is related to context-awareness is the distribution of interfaces (Fig. 1). Modularity has implied the use of different interfaces for different components of the solution. This fact adds another level of customization to different needs and objectives. In addition, one of the interfaces is a distributed user interface, which is the one related to tasks and oriented towards operators. As these users have to manage large quantities of information, the related interface is deployed over three monitors to support three domains for analysis.

3 Conclusions and Future Work

Healthcare environments are an important research area in which to make efforts to improve working conditions. Medical staff have to be able to address complex situations in which mobility, multidisciplinary and dynamic features appear as constraints. Under these conditions, an appropriate way to help users is to offer a system for their daily duties with the capacity to adapt itself according to their needs. For this purpose, this paper has presented Ubi4Health, a comprehensive solution for healthcare environments, focusing on the contribution related to context-awareness. The solution supports users with appropriate and necessary information and functionality on the basis of their current context. In addition, the use of multiple devices, distributed interaction mechanisms and user interfaces has allowed us to go another step forward. A novel scenario appears as Ubi4Health is able to reach an improved level of user experience. Based on the user's context, the solution provides employees and patients with the appropriate device, user interface and way to interact with it at the time.

There are interesting lines of future work connected with the presented context-aware scenario. Currently, the authors are considering the possibility of extending the devices mentioned. For instance, wearable devices offer an interesting way in which Ubi4Health can increase its mobile capabilities and ways of interaction.

Acknowledgements. This work has been partially funded by project with reference 0215ITA017 from the Castilla-La Mancha Regional Government: Junta de Comunidades de Castilla-La Mancha, Consejería de Empleo y Economía.

References

1. Lymberis, A., Smart, A.: Wearables for remote health monitoring, from prevention to rehabilitation: current R&D, future challenges. In: Proceedings of the 4th International IEEE EMBS Conference on Information Technology Applications in Biomedicine, pp. 272–275 (2003)
2. Bardram, J.E., Bossen, C.: Moving to get aHead: local mobility and collaborative work. In: Proceedings of CSCW, pp. 355–374 (2003)
3. Bardram, J.E., Christensen, H.B.: Pervasive computing support for hospitals: an overview of the activity-based computing project. Pervasive Comput. **6**(1), 44–51 (2007)
4. Bricon-Souf, N., Newman, C.R.: Context-awareness in health care: a review. Int. J. Med. Inform. **76**, 2–12 (2007)
5. Weiser, M.: The computer for the twenty-first century. Sci. Am. **265**(3), 94–104 (1991)
6. Favela, J., Martínez, A.I., Rodríguez, M.D., González, V.M.: Ambient computing research for healthcare challenges, opportunities and experiences. Computación y Sistemas **12**(1), 109–127 (2008)
7. Garrido, J.E., Penichet, V.M., Lozano, M.D.: Ubi4Health: ubiquitous system to improve the management of healthcare activities. In: Proceedings of Pervasive 2012 Conference, ACM Digital Library, Newcastle, UK (2012)
8. Gonzales, A.L., Pollak, J.P., Retelny, D., Baumer, E.P., Gay, G.: A mobile application for improving health attitudes: being social matters. In: CHI 2011, Vancouver, BC, Canada, 7–12 May 2011
9. Bardram, J.E., Hansen, T.R., Mogensen, M., Soegaard, M.: Experiences from real-world deployment of context-aware technologies in a hospital environment. In: Dourish, P., Friday, A. (eds.) UbiComp 2006. LNCS, vol. 4206, pp. 369–386. Springer, Heidelberg (2006)
10. Kjeldskov, J., Skov, M.B.: Supporting work activities in healthcare by mobile electronic patient records. In: Masoodian, M., Jones, S., Rogers, B. (eds.) APCHI 2004. LNCS, vol. 3101, pp. 191–200. Springer, Heidelberg (2004)
11. Garrido, J.E., Penichet, V.M., Lozano, M.D., Quigley, A., Kristensson, P.O.: AwToolkit: attention-aware user interface widgets. In: Proceedings of the 2014 International Working Conference on Advanced Visual Interfaces 2014. ACM, Como (2014)

Distributed Tabletops: Study Involving Two RFID Tabletops with Generic Tangible Objects

Amira Bouabid[1,2], Sophie Lepreux[1], and Christophe Kolski[1(✉)]

[1] LAMIH-UMR CNRS 8201, University of Valenciennes, Valenciennes, France
{amira.bouabid,sophie.lepreux,
christophe.kolski}@univ-valenciennes.fr
[2] SETIT, University of Sfax, Sfax, Tunisia
amira.bouabid@gmail.com

Abstract. This paper describes a study on an innovative system designed to support remote collaborative. This system is a game running on tabletops with tangible interaction. Twelve test groups, each composed of three participants, tested a distributed application for learning and recognition of colors. We propose a set of generic tangible objects. They model a set of collaborative styles which are possible between users. Our goal is to obtain objects that provide remote collaboration among users of such interactive tabletops. This study is supported by observations, trace analysis and questionnaires. In this study, we analyze if the use of generic objects is easy and understandable by users in case of remote collaboration. The user satisfaction when using the distributed tangible tabletops is also studied.

Keywords: Tangible interaction · Tabletop · Distributed UI · Remote collaboration · Tangible object · Tangiget · RFID

1 Introduction

Collaborative work proves important within a team, or more generally within a group of users. In fact, team people often needs to exchange ideas [9], to work on common tasks [10] or to be informed about the progress of a task [11]. Much work has been done on the subject concerning the flow of information between users and platforms [2]. We seek in this work to facilitate the collaboration between different people working together. Our objective is to propose a system that provides remote collaboration throw interactive tabletops with tangible interaction [1]. Collaboration in this system is based on a set of generic tangible objects, called tangigets, initially defined by Lepreux *et al.* in [8] and Caelen and Perrot in [3]. These tangigets will materialize a set of collaboration styles listed by Isenberg et al. in [4].

In the paper we describe the design of the system and the interaction, as well as principles for remote collaboration on tangible tabletops. The study is then presented, and the first results are explained. Finally the paper concludes and proposes research perspectives.

S. Casteleyn et al. (Eds.): ICWE 2016 Workshops, LNCS 9881, pp. 167–173, 2016.
DOI: 10.1007/978-3-319-46963-8_15

2   System Design: DUI on Two RFID Tabletops

As Distributed User Interface (DUI), we use two *TangiSense* interactive tabletops allowing tangible interaction [6] (designed by the RFidees Company; see www.rfidees.fr). These tabletops use the RFID technology to recognize objects placed on the surface, as shown in Fig. 1.

Fig. 1. Two different user interfaces shown on the interactive tabletops with some of used (business and generic) tangible objects: on the left, user Interface displayed on the *child* tabletop during the correction of exercise; on the right, interface displayed on the *adult* tabletop during the supervision of the exercise realization

The application used in our study is a distributed version of an application on RFID tabletop allowing the learning and recognition of colors; the initial version was presented in [7]. A set of business objects is used; they present various colorless pictures. For this application the final users are very young children learning colors; they have to arrange the business objects in color areas according to dominant colors. These objects are divided into 4 categories; each category contains 8 objects of varying difficulty levels.

A set of generic tangible objects, called tangigets, is used to ensure remote collaboration between interconnected tabletops. These objects give users the ability to interact remotely using the features provided by the interactive tabletop.

3   Interaction Design: Use of Generic Tangible Objects

Remote collaboration between users on tabletops is carried out by the use of a set of generic objects (tangigets). The aim of these generic objects is to support an inter-user dialogue using only features offered by interactive tabletops:

- *Identification* tangiget: Used to identify users who are currently using the collaborative application and want to enter in collaboration with other users.
- *Task assignment* tangiget: Used to organize tasks between different users of the collaborative application.
- *Starting synchronization* tangiget: Used to synchronize the start of the activity distributed on connected tabletops.
- *Display Mode* tangiget: Used to change the display of the main interface according to the user needs.
- *Request help* tangiget: Used to ask for help or ask a question about a step or a detail of the collaborative activity.
- *Provide help* tangiget: Used to offer help about a step or detail following a request.
- *End task* tangiget: Used to mark the end of a task and/or to switch to another task.
- *Criticism* tangiget: Used to work on all of the activity (not on one task).

Figure 2 shows two examples of Tangigets used in this application: they correspond respectively to *Criticism* tangiget (*Magician* object) and *Provide help* tangiget (*Erase* object).

Fig. 2. Some of used tangigets in the application: on the left, *Provide help* tangiget (represented by *Erase* object); on the right, *Criticism* tangiget (represented by *Magician* object)

4 Remote Collaboration on Tangible Tabletops

Isenberg et al. [4] listed eight collaboration styles covering any type of collaboration (co-located or remote). The proposed tangigets have to comply with the standards of collaboration. So we instantiated each style by an action on the system by using: (1) one tangiget or (2) a coupling between two tangigets.

Table 1 shows the correspondence between the action of each tangiget and one of the collaboration styles.

Table 1. Tangigets contributing to collaboration styles.

Collaboration styles proposed in [4]	Representative action on the application by the use of a generic object
Active **dis**cussion	*Request help*
View Engaged	*Task assignment*
Sharing of the Same View	*Starting synchronization*
Sharing of the Same Information but using Different Views	*Display mode*
	Identification
Working on the Same Specific Problem	*Task assignment coupled with identification*
Working on the Same General Problem	*Provide help*
Working on Different Problems	*Criticism*
Disengaged	*End task*

5 Case Study

In the application, we propose to use a set of tangigets useful for remote collaboration. Table 2 shows those objects and their main functionalities in the application.

Table 2. Instances of tangigets used for the application.

Type of tangiget	Instantiation	Definition of its role
Identification	Identification	Used to identify the person present remotely and ready to play
Starting Synchronization	Start	Start the game on the two connected tabletops
Task assignment	Category	Assign a category of object to the person identified
Request help	Collaboration area	Request remote assistance by placing in this area object(s) requiring help
Provide help	Erase	Offer help by crossing color areas
Display mode	Focus	Display the results of the exercise (textual representation)
End task	End exercise	Indicate the end of the exercise for each user
Criticism	Magician	Correct remotely the exercise

A presentation of the system and its functioning as well as the functioning of each tangiget was given for each group of three participants. It was followed by a familiarization phase with the interactive tabletops during which the participants were encouraged to try the application and all items offered and to freely ask questions. After this phase of familiarization with the system, tests were started with different scenarios provided. We designed three conditions that varied aspects of the use of the *Help request* and the correction of the exercise. In these different conditions, instructions were provided to users of the *child* tabletop. We aimed to get a definite number of mistakes and requests for help. Figure 3 provides an illustration of the participants of a test group set in relation to their different roles.

Fig. 3. A participant testing the application (*adult* tabletop)

To simulate a remote collaboration, participants were in the same room; the two tabletops were separated by a folding screen to prevent users from each tabletop from seeing the contents of the other tabletop. Moreover, to prevent that the *Parent* participant from being disturbed by possible natural discussions coming from the *child* tabletop, he/she had a headset with music.

After the study, each participant had to complete a questionnaire. The questionnaire firstly concerned information on the usability aspects and participant satisfaction with the system. Secondly, he or she had to fill more specific information on generic objects, ease of use and their significance in relation to their role set by the designer.

Finally, the evaluator used a trace file in which all the games played by the group were recorded in order to understand how they had addressed the problem and to get their reactions on the technologies and principles used. The analysis of the trace file is based mainly on a set of reactions of the user following the action by the other user of the remote tabletop. According to the response received, we classified them into three categories: expected answer, acceptable answer and incoherent answer.

As an example, we illustrate our analyses with two tangigets (1) *Erase* object used by the user of the *adult* tabletop and (2) *Identification* object used by the user of the *child* tabletop. The results of the questionnaires for *Identification* and *Erase* objects are summarized in Fig. 4; the score of each answer is shown for each question. We can find from the figure that participants give high marks on the global situation. They think these tangigets are easy to use and have a meaningful form. Also they have understood the goal of collaboration. To study if the use of tangigets by the participants confirms or not their subjective answers, we analyzed all the trace files in which we recorded the events concerning all games played. We extracted all uses of tangigets; after that we classified them as expected use, acceptable use or incoherent use. Analyses relative to *Identification* and *erase* objects are summarized in Fig. 5.

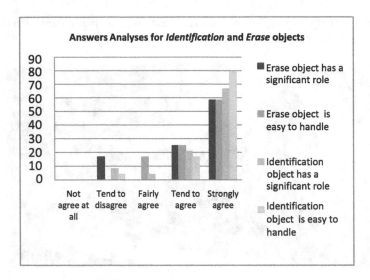

Fig. 4. Subjective answers of participants who used *Identification* and *Erase* tangigets

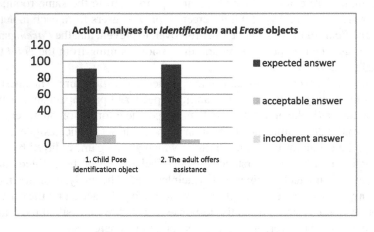

Fig. 5. Objective answers of participants who used *Identification* and *Erase* tangigets

6   Conclusion

In this paper, we introduce an innovative distributed application for the learning and recognition of colors. Generic tangible objects, called tangigets, are used to facilitate collaboration and exchange between distant participants about the exercise. This system takes advantage of large-scale tangible tabletops, (1) providing a simple user interface easy to manipulate; (2) enabling several users to collaborate remotely in each step of the exercise; (3) providing the possibility to cover a set of collaboration styles (in the sense of Isenberg et al. [4]) by the use of tangigets. A study was conducted with twelve groups of three users. The results are promising and show the interest of the distributed approach. In future works we aim to test such tangigets with other more complex applications to verify if collaboration remains easy/possible.

Acknowledgements. The authors thank warmly Sebastien Kubicki for his work on the first centralized version of the application [5]. They thank also the 36 participants and Steve Gabet for his contribution to the distributed version.

References

1. Bouabid, A., Lepreux, S., Kolski, C., Havrez, C.: Context-sensitive and collaborative application for distributed user interfaces on tabletops. In: Proceedings of the 2014 Workshop on Distributed User Interfaces and Multimodal Interaction, pp. 23–26. ACM, New York (2014)
2. Brave, S., Ishii, H., Dahley, A.: Tangible interfaces for remote collaboration and communication. In: Proceedings of the 1998 ACM Conference on Computer Supported Cooperative Work, pp. 169–178. ACM (1998)
3. Caelen, J., Perrot, C.: Bibliothèque d'objets. Deliverable, IMAGIT ANR project. LIG, Grenoble (2011)
4. Isenberg, P., Fisher, D., Morris, M. r., Inkpen, K., Czerwinski, M., MacEachren, A., Miksch, S.: An exploratory study of co-located collaborative visual analytics around a tabletop display. Presented at the 2010 Proceedings of IEEE Symposium on Visual Analytics Science and Technology (VAST 2010), 1 January 2010
5. Kubicki, S., Lepreux, S., Kolski, C.: Evaluation of an interactive table with tangible objects: application with children in a classroom. In: Proceedings of the 2nd Workshop on Child Computer Interaction "UI Technologies and Educational Pedagogy", Conjunction with CHI 2011 Conference (2011)
6. Kubicki, S., Lepreux, S., Kolski, C.: RFID-driven situation awareness on TangiSense, a table interacting with tangible objects. Pers. Ubiquit. Comput. 16(8), 1079–1094 (2011)
7. Kubicki, S., Wolff, M., Lepreux, S., Kolski, C.: RFID interactive tabletop application with tangible objects: exploratory study to observe young children' behaviors. Pers. Ubiquit. Comput. 19(8), 1259–1274 (2015)
8. Lepreux, S., Kubicki, S., Kolski, C., Caelen, J.: From centralized interactive tabletops to distributed surfaces: the tangiget concept. Intl. J. Hum.-Comput. Interact. 28(11), 709–721 (2012)
9. Qin, Y., Liu, J., Wu, C., Shi, Y.: uEmergency: a collaborative system for emergency management on very large tabletop. In: Proceedings of the 2012 ACM International Conference on Interactive Tabletops and Surfaces, pp. 399–402. ACM, New York (2012)
10. Robinson, P., Tuddenham, P.: Distributed tabletops: supporting remote and mixed- presence tabletop collaboration. In: Second Annual IEEE International Workshop on Horizontal Interactive Human-Computer Systems, 2007. TABLETOP 2007, pp. 19–26 (2007)
11. Yamashita, N., Kaji, K., Kuzuoka, H., Hirata, K.: Improving visibility of remote gestures in distributed tabletop collaboration. In: Proceedings of the ACM 2011 Conference on Computer Supported Cooperative Work, pp. 95–104. ACM, New York (2011)

Distributing Interaction in Responsive Cross-Device Applications

Felix Albertos-Marco[✉], Victor M.R. Penichet, and Jose A. Gallud

Computer System Department,
University of Castilla-La Mancha, Albacete, Spain
{felix.albertos,victor.penichet,jose.gallud}@uclm.es

Abstract. With the emergence of the Internet of Things there are many applications in which the interaction is distributed over multiple devices. Developing applications in these scenarios is challenging because there is not enough knowledge and even less consensus on how to distribute interaction. But following the ongoing trends such as responsive web design, why not enable applications to seamlessly adapt interaction to take advantage of available devices at any moment? In this work we introduce the foundations of a new approach called Responsive Cross-Device Applications (RCDA). RCDA applies the idea of responsive Web applications distributing user interactions over the new cross-device ecosystem, taking into account the interactive capacities of devices and users.

Keywords: Responsive interaction · Cross-device applications · Distributed interaction · Interactive capacities

1 Introduction

Distributed Interaction refers to the interaction process in which the user/s use/s dynamically separated input/output devices (e.g. a document viewer [5,6], a snake-like game [5], exploring a distributed map [2,6] or video streaming applications [2,5,6]). With the emergence of the Internet of Things, there is the prospect of a future in which devices can always be connected. Developing applications for these scenarios is challenging because there is not enough knowledge and even less consensus on how to distribute interaction.

In these scenarios, Santosa [4] identified specialization as the main characteristic to improve multi-device interaction and the coordination of devices. Hamilton et al. [2] found that little attention has been paid to interactions which enable the chaining of device functionality, and cross-device experience was found to be painfully absent in today's technologies. Yang presented Panelrama [6], a web-based framework for the construction of applications using distributed user interfaces. It allows users to interact with a single application from multiple devices that dynamically change user interface allocation to best-fit devices. Jokela [3] found that users want to use their devices to seamlessly work with each other, but in practice they continuously encountered problems in multi-device use. Schreiner presented Connichiwa [5], a framework for the development

© Springer International Publishing AG 2016
S. Casteleyn et al. (Eds.): ICWE 2016 Workshops, LNCS 9881, pp. 174–178, 2016.
DOI: 10.1007/978-3-319-46963-8_16

of cross-device web applications focused on the integration of existent devices, independent from the network infrastructure, and offering versatility over application scenarios and API usability. Chi et al. presented Weave [1], a framework for developers to create cross-device wearable interaction by scripting.

Most of the developments are ad-hoc solutions based on particular case studies and/or developers' intuition. They do not fully take into account the whole interactive system: users, devices and tasks. There is a lack of consensus about how to distribute interaction according to the capabilities of the interactive system and their relation to the main goal of every interactive system: task accomplishment. However, following the ongoing trends such as responsive web design, why not enable applications to seamlessly adapt interaction to take advantage of available devices at any moment? This work provides the foundations for a new approach aimed at dealing with the design and development of Responsive Cross-Device Applications (RCDA) in current cross-device applications. The idea applies responsive Web applications to distributing user interactions over the new cross-device ecosystem.

The paper is structured as follows. First, the two pillars of RCDA are presented. Then the foundations of our proposal are described: user and device characterization, adaptation responsibility, runtime adaptations and user patterns. Finally, the conclusions and future work are presented.

2   Responsive Cross-Device Applications

The new approach of RCDA has two pillars: Responsive Web applications, and Cross-Device applications. The following paragraphs explain these two foundational ideas. Nowadays, and even more in the future, it seems to be clear that applications must support a user's tasks over multiple devices. This tendency is somewhat similar to the Web and how Web application development and use evolved to be what they are now: Responsive Web applications. They adapt to some characteristics of devices such as size, screen resolution, orientation or input modalities. But today's environments are far more complicated. In a similar way, the coordination of tasks in cross-device applications is complex [1]. Current approaches do not allow for simultaneous use of devices or cross-device interaction. Suitable applications are needed to deal with the wide range of existent cross-device scenarios [5].

However, responsive design in cross-device applications should also consider the fact that applications distribute functionality among different devices. Therefore, new features regarding such a distribution should be applied in the design. Thus, following the aforementioned two pillars, we propose a new approach for the design of cross-device applications: Responsive Cross-Device Applications (RCDA). RCDA adapts the interaction according to the ecosystem, which is composed of devices, users and tasks. The main goal of RCDA is to support users' tasks in cross-device environments, adapting interaction to facilitate user task completion.

From the analysis of previous works we set the foundations for the design and development of RCDA. The foundations are the following:

1. User and device characterization.
2. Adaptation responsibility.
3. Runtime adaptations.
4. User patterns.

These foundations are described in the following sections.

2.1 Users and Devices Characterization

User and device characterization has to take into account not only properties such as size and display resolution, but also the interactive capabilities of users and devices, the allocation of specific roles to each device and/or actor, the coordination of devices for simultaneous use and what the device is best suited for. In previous approaches, the description and categorization of interactive capabilities has been limited to physical properties such as size and display resolution. Only a few works [6] have used other device capabilities. But these descriptions are limited to specific parts of the user interface and some characteristics of a device's capabilities. In our approach the description of the interactive system is based on the interactive capabilities of the actors involved in the interaction process: users and devices. Therefore, in RCDA the actors have to be described according to their input and output interactive capabilities. These capabilities are mapped to support task completion on the interactive system. This mapping takes into account the description of the current task in terms of interactive capabilities. As a result, the interaction adapts to the task performed by the interactive system and the interactive capabilities of users and devices. The process of mapping interaction capabilities to tasks is still an open topic. But previous approaches can be used, such as defining roles [4] or defining fitness values [6], among others.

2.2 Adaptations Responsibility

The adaptation of user interaction is the responsibility of the system. It is the applications and not the users that have to adapt and move interaction across multiple devices to help users to fulfil their tasks. They are in charge of mapping the interactive capabilities of both users and devices. And, under specific circumstances, users will be able to decide from among multiple interaction paths. Depending on how users are able to decide from among multiple interaction paths, we have set three schemas:

1. Fixed. Users do not decide how the interaction is distributed.
2. Restricted. Users will be able to decide how the interaction is distributed in a limited way. They do not have full control over how to distribute interaction.
3. Free. Users are fully capable of distributing interaction over the interactive system.

Each one of these schemas has its strengths and weaknesses. While the first schema (fixed) is the most restrictive, not allowing users to decide how inter- action is distributed, it is also the most simple from the point of view of the users. They only have to use the system. They are not aware of when, why and how adaptations are made. On the other hand, the free schema is the most complicated for users. They have to decide when and how the distribution has to be made. Therefore, it is important to support the schema that best fits the interactive system requirements.

2.3   Runtime Adaptations

RCDA has to dynamically categorize devices and users to change interaction to best-fit users' tasks completion. It is not enough to describe an application's scenario statically. Multi-device scenarios tend to change their configuration, adding or removing devices/users and consequently changing interaction capa- bilities. RCDA has to be aware of these circumstances. Therefore, they have to be up to date with the interactive input and output capabilities of the actors involved, adapting interaction in runtime. Therefore, RCDA has to support flexi- ble mapping between these capabilities and the task the user/s is/are performing.

2.4   User Patterns

RCDA has to identify and support users' patterns for task execution in cross- device scenarios. Depending on how the task is transferred between devices, the information moves between them or their physical or logical arrangement in patterns that are classified as serial or parallel. In the serial or sequential pattern the task is transferred from one device to another [3,4]. One specialization of this pattern is the producer-consumer pattern, in which the content is produced on one device for it then to be moved to, and used by, another device [4]. In the parallel pattern many devices are coordinated for simultaneous use [4]. In the performer-informer pattern, one device has the resources to support the creation of content on other device [4]. In the performer-informer-helper pattern users use support devices (helpers) to perform traversal tasks [4]. In the controller- viewer/analyzer pattern users wish to combine device functionalities to perform different aspects of a single primary task [4]. Users also perform resource lending (borrowing resources from other devices), related parallel use (all devices are involved in a single task) and unrelated parallel use (devices are involved in different tasks) [3]. Devices can also be physically organized by mapping their physical position within the environment, following the spatial mapping pattern [2]. But there may be no spatial mapping when organizing devices [2]. The devices = windows pattern is used when devices are used as windows in the WIMP paradigm [2]. In the device = role pattern each devices is mapped with a single role for the current task [2]. In the device = storage pattern each device shows, for example, a specific document for the task [2].

3   Conclusions and Future Work

In this paper we have introduced the foundations of a new approach to handling the distribution of interactions in cross-device applications, and we have called it Responsive Cross-Device Applications (RCDA). The main characteristics of this approach are: it goes beyond screen size to design adaptations; applications are in charge of adapting interaction in cross-device environments, not users; and it makes this happen dynamically. Also, depending on how the task is transferred between devices, how the information moves between them or their physical or logical arrangement, the patterns presented have to be taken into account for the development of RCDA. As future work, and due to the complexity of RCDA, we will continue with the development of tools for the description and simulation of such complex scenarios.

Acknowledgments. This work has been partially supported by the fellowship 2014/10340 from the University of Castilla-La Mancha.

References

1. Chi, P., Li, Y.: Weave: scripting cross-device wearable interaction. In: Proceedings of the 33rd Annual ACM Conference on Human Factors in Computing Systems (CHI 2015), pp. 3923–3932. ACM, New York (2015). doi:10.1145/2702123.2702451
2. Hamilton, P., Wigdor, D.J.: Conductor: enabling and understanding cross-device interaction. In: Proceedings of the SIGCHI Conference on Human Factors in Computing Systems (CHI 2014), pp. 2773–2782. ACM, New York (2014). doi:10.1145/2556288.2557170
3. Jokela, T., Ojala, J., Olsson, T.: A diary study on combining multiple information devices in everyday activities and tasks. In: Proceedings of the 33rd Annual ACM Conference on Human Factors in Computing Systems, CHI 2015, pp. 3903–3912. ACM, New York (2015). ISBN 978-1-4503-3145-6
4. Santosa, S., Wigdor, D.: A field study of multi-device workflows in distributed workspaces. In: Proceedings of the ACM International Joint Conference on Pervasive and Ubiquitous Computing (UbiComp 2013), pp. 63–72. ACM, New York (2013). doi:10.1145/2493432.2493476
5. Schreiner, M., Radle, M., Jetter, H., Reiterer, H.: Connichiwa: a framework for cross-device web applications. In: Proceedings of the 33rd Annual ACM Conference Extended Abstracts on Human Factors in Computing Systems (CHI EA 2015), pp. 2163–2168. ACM, New York (2015). doi:10.1145/2702613.2732909
6. Yang, J., Wigdor, D.: Panelrama: enabling easy specification of cross-device web applications. In: Proceedings of the SIGCHI Conference on Human Factors in Computing Systems (CHI 2014), pp. 2783–2792. ACM, New York (2014). doi:10.1145/2556288.2557199

Towards User-Defined Cross-Device Interaction

Audrey Sanctorum[(⊠)] and Beat Signer

Web & Information Systems Engineering Lab,
Vrije Universiteit Brussel, Pleinlaan 2, 1050 Brussels, Belgium
{asanctor,bsigner}@vub.ac.be

Abstract. Over the last decade we have seen various research on distributed user interfaces (DUIs). We provide an overview of existing DUI approaches and classify the different solutions based on the granularity of the distributed UI components, location constraints as well as their support for the distribution of state. We propose an approach for user-defined cross-device interaction where users can author their customised user interfaces based on a hypermedia metamodel and the concept of active components. Furthermore, we discuss the configuration and sharing of customised distributed user interfaces by end users where the focus is on an authoring rather than programming approach.

Keywords: Cross-device interaction · DUIs · End-user development

1 Classification of Distributed User Interfaces

Over the last few decades, distributed user interfaces (DUIs) have gained a lot of attention [23]. Various terms have been introduced in order to differentiate between different DUI systems, ranging from multi-device and multi-display interaction to interactive spaces and cross-device interaction. Multi-device applications started to emerge already in the late twentieth century when, for example, Robertson et al. [26] presented a system that allowed users to interact with a TV by using a personal digital assistant (PDA) and a stylus. A limitation of this early system was that the information flow was only possible in one direction from the PDA to the TV, which prevented a user from capturing information from the TV to their PDA. Only a year later, Rekimoto [25] introduced the pick-and-drop technique which allowed users to exchange information in any direction by picking up an object on one computer screen and dropping it on another screen by using a digitiser stylus. Since then, research in cross-device interaction has gone a long way and different techniques, interaction possibilities, frameworks and applications have been developed [10]. In order to pass information across devices, Frosini et al. [12] make use of QR codes. The Deep Shot system [7] supports information sharing between a smartphone and a computer screen via the smartphone's camera. While Deep Shot uses a feature matching algorithm, in the Conductor system Hamilton and Wigdor [13] proposed another solution which enables the distribution of information across different

© Springer International Publishing AG 2016
S. Casteleyn et al. (Eds.): ICWE 2016 Workshops, LNCS 9881, pp. 179–187, 2016.
DOI: 10.1007/978-3-319-46963-8_17

tablets via broadcasting. An alternative way for cross-device communication has been presented by Rädle et al. [24] with the HuddleLamp system which uses a lamp with an integrated camera to track hand movements and the position of any mobile device that has been placed on the table the lamp is standing on. While HuddleLamp covers a limited area, other systems allow for interactions across a room [5,11,16,18,31], a network environment [6,7,13–15,21,32] or have no space limitations and can be used anywhere [2–4,8,12,17,20,27].

Apart from the 'space' dimension, other criteria can be used to highlight the differences between existing DUI solutions. For example, the granularity of the distributed components is another interesting criteria. Some systems, such as the ARIS system [5] where application windows can be moved across devices, only support the distribution of applications as a whole. Other solutions offer a specific set of components that can be distributed. For example, MultiMasher [15] supports the distribution of arbitrary elements from any website while Melchior et al. [20] support the distribution of application widgets and arbitrary pixels on a screen. This also offers the possibility to transfer the state of an application across devices and to have synchronous views of the shared data. The distribution of user interfaces and user interface components is often performed via a message passing mechanisms [1,3,4,13,16,18,21,27]. Moreover, certain systems developed their own software infrastructure for the sharing of information across devices as seen with BEACH in the i-LAND [31] project.

Distributed user interfaces are often built following a client-server architecture, where the server plays a central role in keeping each user interface on the different devices up to date [6,7,13–15,21,32]. However, this approach limits the interaction space of the connected devices since all devices need to be connected to the server. A solution to overcome this limitation is the use of peer-to-peer networks without a need for a central server [3,4,20].

While Demeure et al. [9] proposed a reference framework to differentiate existing DUI approaches based on the four dimensions of computation, communication, coordination and configuration, in Fig. 1 we provide our classification of the discussed approaches based on their constraints in terms of location and the supported granularity for distributed UI components. On the x-axis we go from local solutions on the left to solutions without any space limitation on the right. On the y-axis we differentiate between systems where the entire UI and data can be shared to approaches that support the sharing of UI elements at a finer granularity. Note that we further highlight systems that support the distribution of state by labelling them in a bold blue font.

Some of the previously described systems focus on the portability, others centre around the collaborative aspect and a third group focusses on making it easier for designers and developers to create DUI applications. However, almost none of them deals with making the frameworks or applications available to end users without programming skills. Certain systems like Weave provide "easy-to-use" scripting languages to build DUIs or to ease the distribution across different devices. WebSplitter [14] provides users with an XML file which contains the distribution of the UI elements across devices. Going a step further,

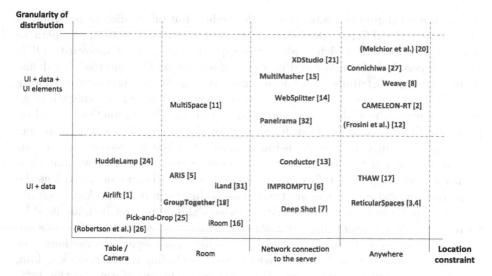

Fig. 1. Classification of DUIs based on location constraints (x-axis) and the supported granularity of distribution (y-axis) (Color figure online)

XDStudio [21] provides a web-based authoring environment for designers who have only basic web development experience. Finally, Husmann et al. [15] presented MultiMasher, a tool for technical as well as non-technical users. Multi-Masher is limited to the distribution of website components and users cannot distribute their applications and data. These systems make a step into the right direction but often still represent "closed solutions" where it is up to the developer or designer to define how exactly an end user can interact across devices.

2 Proposed Approach

In order to overcome some of the shortcomings of existing distributed user interface approaches described in the previous section, we aim at empowering end users to create, modify and reconfigure DUIs. This allows end users to combine multiple interfaces and build their own customised distributed user interfaces in order to better support their daily activities. A first question that arises in this context is how to concretely enable end users to define their customised interactions across electronic devices dealing with digital information and services. Further questions are: *"What will end users be able to modify?"*, *"How much control will end users have in terms of the granularity of the UI components to be distributed?"*, *"Will end users be limited by a specific location, space or office setting?"*, *"Will end users be able to share their configuration of customised UIs?"* and *"Can end users reuse parts of other configurations?"*.

In order to allow end users to customise existing user interfaces as well as to define their own new distributed interfaces, there is a need for end-user authoring tools that enable the specification of cross-device interactions. Note that the authoring should not rely on a single method but offer different possibilities for unifying the different devices forming part of the interaction. We plan to develop a framework which enables the rapid prototyping of innovative DUIs by developers but also allow end users to customise existing interfaces or define their own new distributed user interfaces via a dedicated end-user authoring tool. However, we would also like to investigate new forms of authoring which go beyond the graphical definition and composition of DUI interactions based on programming-by-example. In order to develop such a rapid prototyping framework, we are currently investigating a model and the necessary abstractions for the end-user definition of cross-device interactions. Thereby, we aim for a solution where digital interface components, tangible UI elements as well as the triggered application services are treated as modular components. Any programming efforts for new cross-device user interfaces should further be minimised by turning the development into an *authoring rather than a programming activity*.

A number of authors presented models for cross-device interactions. For example, Nebeling et al. [22] introduced a model including the user, device, data, private session and session concepts that are used in their platform. Another platform model which is more centered around the concrete distribution of UIs has been introduced by Melchior [19]. In their case, a platform has the three main component categories of connection, hardware and audio/video. Existing models are often designed for a specific platform or system introduced by the authors, focus on the distribution across devices and lack concepts such as the re-usability of UI components, the different classes of users as well as the sharing of DUI configurations between users. We are currently developing a more user-centric cross-device interaction model addressing some of these issues.

A promising approach that we are currently investigating for modelling loosely coupled interaction between user interface components and various forms of actions is presented in the work of Signer and Norrie [28]. They proposed a resource-selector-link (RSL) hypermedia metamodel which enables the linking of arbitrary digital and physical entities via a resource plug-in mechanism. In our context, resources can be seen as different user interface components which can be linked together. Since often we do not want to link an entire resource but only specific parts of a resource, such as parts of a UI in order to control the granularity of the distributed user interface components as discussed in the previous section. The concept of a selector allows us to address parts of a specific resource. A detailed description of the RSL hypermedia metamodel can be found in [28]. An important RSL concept for realising our goal of DUI state transfer and the execution of third-party application logic is the concept of so-called active components [29,30]. An active component is a special type of resource representing a piece of program code that gets executed once a link to an active component is followed. This has the advantage that one can trigger some application logic by simply linking the UI, or parts of the UI represented by resources and selectors, to an active component. More importantly, an active component does not have to implement the application logic itself but can also act as a proxy for functionality offered by any third-party application. We foresee that the concept of active components can enable the rapid prototyping of cross-device applications by simply defining links between the necessary components. We further plan to address a number of other issues such as how to clearly separate the cross-device interactions from the underlying shared data and application state, different forms of lightweight data exchange between devices as well as the possibility for configuring interactions in an ad-hoc manner.

In addition to our model for cross-device interaction, we are currently designing an architecture and implementation of a framework which provides the necessary functionality to communicate between different user interface components and the corresponding application services. In order to facilitate replication and to enable the synchronisation of UIs and UI components on different devices, a distributed model-view-controller (dMVC) pattern, which has proven to be efficient by Bardam et al. [3,4], might be used. Another possibility is to follow the replication-based model of Biehl et al. [6] which captures the application window's pixel data and reproduces it on other devices. On the implementation level, we plan to use an event-based system and a publish/subscribe message passing mechanism as used by other systems [1,3,4,7,8,13,16,18,21,27]. Since we aim for a portable solution that can be used at any location without prior installation, such as some of the systems discussed earlier in the previous section [2,8,12,17,20,27], we consider using JavaScript to support the distribution across devices as seen in other DUI systems [7,8,15,21,24,27].

While we plan to base our cross-device interaction model on some of the concepts introduced in the RSL metamodel, we also intend to develop a framework providing the necessary functionality to communicate between different user interface components as well as a mechanism to discover and manage existing user interface components (e.g. resource/selector plug-ins and active components). The latter is encapsulated in the Developer Registry component shown in the general architecture overview of our envisioned framework in Fig. 2. The Active Components sub-component stores all active components that have been implemented by developers while the Resource/Selector Plug-ins sub-component stores all the resource and selector plug-ins. In addition, it is essential to have a user interface registration and discovery service where end users can upload their newly composed interfaces to share them with other users. This functionality is encapsulated in the End User Registry component. The registry service is used to keep track of the different UI components in a given setting by means of user profiles. If a user created some new cross-device interactions between different UI components, they might be interested to make their new DUI configuration available to other users in a similar way as developers do this in the first place. For this purpose, users can post their own cross-device configurations to the Configuration Pool. This has the advantage that the interactions can be adapted and modified by different users over time which might be seen as an evolutionary development of the corresponding interactions. While in most cases users will adapt existing solutions based on their individual preferences, it can of course also be interesting to see whether some general interaction patterns evolve over time.

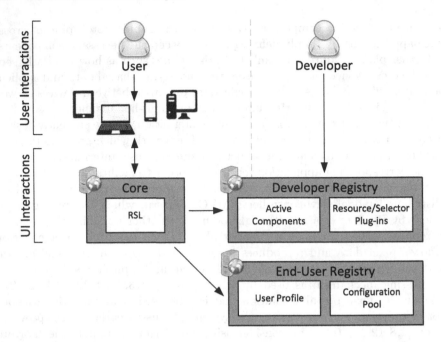

Fig. 2. Architecture for user-defined cross-device interactions

We foresee a synergy between interactions that have been predefined by a developer and are used as is, the ones that are slightly adapted by end users, as well as newly defined interactions by end users. Note that we do not plan to delegate all the interaction definitions to the end user. End users might still mainly rely on predefined interactions but will have the possibility to adapt them or add new cross-device interactions if necessary. By providing the end user the freedom to adapt the interactions, we address the issue that individual users might have slightly different requirements for certain tasks which makes it impossible to design interactions which perfectly suit everybody. Furthermore, the acceptance of specific user interfaces might be increased if end users have the chance to better integrate them with their existing work practices.

A last point that we want to address, which is related to the idea that users can share their interaction components, is how to control the granularity of the shared components. Based on the model that we plan to develop, a user could only share simple user interface components which trigger a single action via an active component. However, a user might often want to share more complex interactions involving multiple devices which can trigger different actions. We will therefore investigate how our model has to be extended in order to group multiple components together and share them as a package. Since the RSL metamodel offers the concept of structural links which can be used to group multiple entities, we plan to initially address this issue by analysing whether and how structural links could be used for defining more complex cross-device interactions by grouping multiple components.

3 Conclusion

We have proposed an approach for user-defined cross-device interaction based on a hypermedia metamodel where UI components can be linked to different application logic at any level of granularity based on the concept of active components. We have further introduced an architecture for the sharing of user-defined user interface components and discussed the authoring of these UIs.

Acknowledgements. The research of Audrey Sanctorum is funded by a PhD grant of the Research Foundation Flanders (FWO).

References

1. Bader, T., Heck, A., Beyerer, J.: Lift-and-drop: crossing boundaries in a multi-display environment by airlift. In: Proceedings of AVI 2010, Roma, Italy, May 2010
2. Balme, L., Demeure, A., Barralon, N., Calvary, G.: CAMELEON-RT: a software architecture reference model for distributed, migratable, and plastic user interfaces. In: Markopoulos, P., Eggen, B., Aarts, E., Crowley, J.L. (eds.) EUSAI 2004. LNCS, vol. 3295, pp. 291–302. Springer, Heidelberg (2004)
3. Bardram, J., Gueddana, S., Houben, S., Nielsen, S.: ReticularSpaces: activity-based computing support for physically distributed and collaborative smart spaces. In: Proceedings of CHI 2012, Austin, USA, May 2012
4. Bardram, J., Houben, S., Nielsen, S., Gueddana, S.: The design and architecture of reticularspaces: an activity-based computing framework for distributed and collaborative smartspaces. In: Proceedings of EICS 2012, Copenhagen, Denmark, June 2012
5. Biehl, J.T., Bailey, B.P.: ARIS: an interface for application relocation in an interactive space. In: Proceedings of GI 2004, London, Canada, May 2004
6. Biehl, J.T., Baker, W.T., Bailey, B.P., Tan, D.S., Inkpen, K.M., Czerwinski, M.: IMPROMPTU: a new interaction framework for supporting collaboration in multiple display environments and its field evaluation for co-located software development. In: Proceedings of CHI 2008, Florence, Italy, April 2008
7. Chang, T., Li, Y.: Deep shot: a framework for migrating tasks across devices using mobile phone cameras. In: Proceedings of CHI 2011, Vancouver, Canada, May 2011
8. Chi, P.P., Li, Y.: Weave: scripting cross-device wearable interaction. In: Proceedings of CHI 2015, Seoul, Republic of Korea, April 2015
9. Demeure, A., Sottet, J., Calvary, G., Coutaz, J., Ganneau, V., Vanderdonckt, J.: The 4C reference model for distributed user interfaces. In: Proceedings of ICAS 2008, Gosier, Guadeloupe, March 2008
10. Elmqvist, N.: Distributed user interfaces: state of the art. In: Distributed User Interfaces: Designing Interfaces for the Distributed Ecosystem. Human-Computer Interaction Series (2011)
11. Everitt, K., Shen, C., Ryall, K., Forlines, C.: MultiSpace: enabling electronic document micro-mobility in table-centric, multi-device environments. In: Proceedings of Tabletop 2006, Adelaide, Australia, January 2006
12. Frosini, L., Manca, M., Paternò, F.: A framework for the development of distributed interactive applications. In: Proceedings of EICS 2013, London, UK, June 2013

13. Hamilton, P., Wigdor, D.J.: Conductor: enabling and understanding cross-device interaction. In: Proceedings of CHI 2014, Toronto, Canada, April 2014
14. Han, R., Perret, V., Naghshineh, M.: WebSplitter: a unified XML framework for multi-device collaborative web browsing. In: Proceedings of CSCW 2000, Philadelphia, USA, December 2000
15. Husmann, M., Nebeling, M., Pongelli, S., Norrie, M.C.: MultiMasher: providing architectural support and visual tools for multi-device mashups. In: Benatallah, B., Bestavros, A., Manolopoulos, Y., Vakali, A., Zhang, Y. (eds.) WISE 2014, Part II. LNCS, vol. 8787, pp. 199–214. Springer, Heidelberg (2014)
16. Johanson, B., Fox, A., Winograd, T.: The interactive workspaces project: experiences with ubiquitous computing rooms. IEEE Pervasive Comput. 1(2), 67–74 (2002)
17. Leigh, S., Schoessler, P., Heibeck, F., Maes, P., Ishii, H.: THAW: tangible interaction with see-through augmentation for smartphones on computer screens. In: Proceedings of TEI 2015, Stanford, CA, USA, January 2015
18. Marquardt, N., Hinckley, K., Greenberg, S.: Cross-device interaction via micromobility and F-formations. In: Proceedings of UIST 2012, Cambridge, USA, October 2012
19. Melchior, J.: Distributed user interfaces in space and time. In: Proceedings of EICS 2011, Pisa, Italy, June 2011
20. Melchior, J., Grolaux, D., Vanderdonckt, J., Roy, P.V.: A toolkit for peer-to-peer distributed user interfaces: concepts, implementation, and applications. In: Proceedings of EICS 2009, Pittsburgh, USA, July 2009
21. Nebeling, M., Mintsi, T., Husmann, M., Norrie, M.C.: Interactive development of cross-device user interfaces. In: Proceedings of CHI 2014, Toronto, Canada, April 2014
22. Nebeling, M., Zimmerli, C., Husmann, M., Simmen, D.E., Norrie, M.C.: Information concepts for cross-device applications. In: Proceedings of DUI 2013, London, UK, June 2013
23. Paternò, F., Santoro, C.: A logical framework for multi-device user interfaces. In: Proceedings of EICS 2012, Copenhagen, Denmark, June 2012
24. Rädle, R., Jetter, H., Marquardt, N., Reiterer, H., Rogers, Y.: HuddleLamp: spatially-aware mobile displays for ad-hoc around-the-table collaboration. In: Proceedings of ITS 2014, Dresden, Germany, November 2014
25. Rekimoto, J.: Pick-and-drop: a direct manipulation technique for multiple computer environments. In: Proceedings of UIST 1997, Banff, Canada, October 1997
26. Robertson, S.P., Wharton, C., Ashworth, C., Franzke, M.: Dual device user interface design: PDAs and interactive television. In: Proceedings of CHI 1996, Vancouver, Canada, April 1996
27. Schreiner, M., Rädle, R., Jetter, H., Reiterer, H.: Connichiwa: a framework for cross-device web applications. In: Proceedings of CHI 2015, Seoul, Republic of Korea, April 2015
28. Signer, B., Norrie, M.C.: As we may link: a general metamodel for hypermedia systems. In: Proceedings of ER 2007, Auckland, New Zealand, November 2007
29. Signer, B., Norrie, M.C.: A framework for developing pervasive cross-media applications based on physical hypermedia and active components. In: Proceedings of ICPCA 2008, Alexandria, Egypt, October 2008

30. Signer, B., Norrie, M.C.: Active components as a method for coupling data and services – a database-driven application development process. In: Norrie, M.C., Grossniklaus, M. (eds.) Object Databases. LNCS, vol. 5936, pp. 59–76. Springer, Heidelberg (2010)
31. Streitz, N.A., Geißler, J., Holmer, T., Konomi, S., Müller-Tomfelde, C., Reischl, W., Rexroth, P., Seitz, P., Steinmetz, R.: i-LAND: an interactive landscape for creativity and innovation. In: Proceedings of the CHI 1999, Pittsburgh, USA, May 1999
32. Yang, J., Wigdor, D.: Panelrama: enabling easy specification of cross-device web applications. In: Proceedings of CHI 2014, Toronto, Canada, April 2014

Optimally Storing the User Interaction in Mashup Interfaces Within a Relational Database

Antonio Jesús Fernández-García[1]([✉]), Luis Iribarne[1],
Antonio Corral[1], Javier Criado[1], and James Z. Wang[2]

[1] Applied Computing Group, University of Almeria, Almería, Spain
{ajfernandez,luis.iribarne,acorral,javi.criado}@ual.es
[2] The Pennsylvania State University, State College, USA
jwang@ist.psu.edu

Abstract. Cross-device applications that have user interfaces managed in multiple forms of interaction are prevalent. In particular, component-based (or *mashup*) applications are growing in popularity due to their easiness to build customized user interfaces with pieces of information from different sources. Since the user interaction on mashup interfaces can generate a large quantity of data, which can be useful to improving the interaction and usefulness of the application, it may involve the creation of cloud infrastructures to manage the dynamic distributed user interfaces within this context. Storing the generated data from the interaction performed over the user interface can be challenging. To achieve these goals, in this paper, a relational database for storing this interaction information generated on distributed user interfaces is proposed. Thus, user interaction over heterogeneous interfaces and devices described in detail, will be easily accessible for further analysis using machine learning and data mining techniques to offer a better user experience.

Keywords: Mashup · User interaction · Multiforms of interaction · Cross-device applications · Relational database

1 Introduction

Today users consume information through heterogeneous devices such as computers, laptops, tablets or smartphones. Moreover, each device has a different way to interact with; some of them support classical forms of interaction by means of keyboard and mouse, others interact through touch interfaces, gestural interfaces or voice recognition (*Natural User Interaction*, NUI). Other interaction technologies are emerging, *e.g.*, virtual reality or wearables & IoT solutions.

Frequently, a same application needs to be available for multiple devices via different user interfaces (UIs). Users expect applications to be accessible via any device regardless of the screen size, the type of interaction, or the technologies involved in it [1]. It becomes even more complicated when it concerns to the user

S. Casteleyn et al. (Eds.): ICWE 2016 Workshops, LNCS 9881, pp. 188–195, 2016.
DOI: 10.1007/978-3-319-46963-8_18

configuration of the interface. Usually, UIs manage some configuration options and remember the behavior and the interactions performed by users and more features progressively.

Due to the increasing amount of services and APIs available, it is becoming a standard practice to use content from many sources in a Web application through a single UI. These UIs, commonly referred to as component-based UIs or *mashup*, allow users to easily customize their UI by employing different pieces of information or data creating their own tailored UI. Mashup interfaces [2] are typically used. Due to their granularity (coarse-grained) they facilitate the adaptation of their internal structure. A cloud infrastructure for the management of mashup UI can be a natural approach. In previous works a series of Web services, located in the platform-independent layer of a cloud infrastructure, have been created to support component-based architectures of mashup UI [3,4]. These services include features such as managing users, component or sessions; and the administration of modules, controllers and databases that underlies below. This infrastructure provides a solid base to create dynamic UIs [5].

This paper focuses on the interaction of users over *mashup* interfaces. There is an extraordinary potential in analyzing the interaction performed in mashup interfaces to improve the user experience by adapting the interface at run-time to the users' requirements and even stepping to the users' needs. Using *machine learning* and *data mining* techniques over the interaction data acquired from users makes it possible to discover behavioral patterns and create prediction models. For that, it is necessary not only to acquire the data but also to know exactly the morphology of component-based UIs and to create an *optimized* relational database that can storage all the data for further analysis. Currently, there is no database schema proposal to store user interaction. The problem is not straightforward because there are many mashup UI and each one of them has a different purpose and their users have different domain knowledge, skills and expectations. Also, Web technologies are diverse and for that reason the data acquisition process that has to be implemented to store the interaction in the database should be independent of the technology used to develop the mashup UI, as well as not intrusive and totally transparent for users.

The rest of the paper is organized as follows. Section 2 describes the morphology of a basic mashup graphic user interface (GUI). Section 3 proposes a relational database to store the interaction produced over this type of interfaces. Section 4 shows a query to the information gathered in the database deployed in a real mashup. We conclude and provide future directions in Sect. 5.

2   Essential Mashup GUI Morphology

Mashup User Interfaces (mashup UI) are Web applications that integrate one or more components from one or more sources to create a unique UI that combines different components that might or might not have relationship among them. This section explains in detail how mashup UIs are composed, with focus on mashup GUIs. A standard interface that covers all the common aspects of mashups has been considered. There are many more features available in specific interfaces but all of them have some core elements and operations, which have been taken into consideration in this morphology definition.

There are many examples of commercial component-based interfaces. Nowadays, mashup interfaces (or component-based interfaces) are widespread in commercial software, particularly in Web applications [6]. Geckoboard is a KPI dashboard surface where users can visualize and work with their most important business data in real-time focusing on sales, marketing or operations among other features [7]. Cyfe allows users to build their own dashboard adding pieces of information through social media, analytics, sales, finance or project management components among others [8]. ENIA (Environmental Information Agent) is a mashup component-based GUI for environmental management used by the Andalusian Environmental Information Network (REDIAM) [9], a public organization that belongs to the Andalusian Regional Government (Spain) [10].

Figure 1 conceptually presents a component-based interface where the elements that form it are shown. Obviously, there may be many more elements and they could be positioned differently. All mashup UIs studied share some core elements and that is what is represented in Fig. 1.

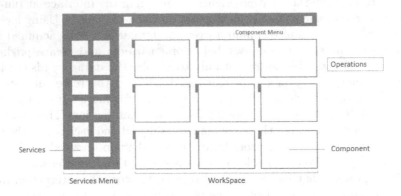

Fig. 1. Conceptual design of a component-based Web application.

Services. The capacities that the mashup application offers. They are available to users in order to operate with them. An instance of a *Service* is a *Component*.

Services menu. In this menu a list of all the *Services* available in the mashup application can be found. Users can navigate through this menu to find the services they may need. Usually, this menu is categorized and grouped by types of services and it has some search tools to locate them directly.

Component. When a user adds a Service to the *workspace* it is automatically transformed into a *Component*. A *Component* is a *Service* that is being used by a user at a certain moment in time. When a *Component* is instantiated, a set of attributes like width, height or position are assigned to it.

Workspace. The *Workspace* is the work area where users have all the *Components* (*Services* instantiated) they are working with.

Operations. All possible actions that are able to be applied over the *Components* such as resize, move or delete, among others.

Therefore, a mashup GUI ($\mathcal{M}$) is defined in the following manner: $\mathcal{M} = \{\mathcal{S}, \overline{\mathcal{S}}, \mathcal{C}, \mathcal{W}, \mathcal{O}\}$. Thus, M is comprised of a set of services $\mathcal{S}$, a service menu $\overline{\mathcal{S}}$, a set of components $\mathcal{C}$, a workspace $\mathcal{W}$ and a set of operations $\mathcal{O}$. The set of *services* $\mathcal{S}$ is defined as $\mathcal{S}=\{S_1, S_2, .., S_N\}$ where N is the number of *services* registered in the information system. The set of *components* $\mathcal{C}$ is defined as $\mathcal{C}=\{C_1, C_2, .., C_L\}$ where L is the number of *components* instantiated in the workspace $\mathcal{W}$. A concrete component C_i has some properties so it could be defined as $C_i = \{PosX, PosY, Width, Height\}$. Finally, the set of operations is defined as $\mathcal{O}=\{Add, Delete, Move, Resize\}$ and they are described below:

Add. Consists in adding a service to the workspace from the services menu, so it is instantiated into a component. When instantiating, some properties such as position in the x-axis, position in the y-axis, width and height are assigned to the component.

Delete. Consists in removing a component from the workspace. That happens mostly because it is of no use and users decide to dispense of it.

Resize. Consists in changing the size assigned to a component. It modifies the 'width' (w) and 'height' (h) properties. Sometimes the Resize operation can be decomposed in several operations such as:

$$Resize(x) - \begin{cases} x = ResizeBigger \mid (w_i * h_i) < (w_{i+1} * h_{i+1}) \\ x = ResizeSmaller \mid (w_i * h_i) > (w_{i+1} * h_{i+1}) \\ x = ResizeShape \mid (w_i * h_i) = (w_{i+1} * h_{i+1}) \\ \qquad \wedge ((w_i \neq w_{i+1}) \vee (h_i \neq h_{i+1})) \end{cases} ,$$

where ResizeBigger operation is considered when the area covered after the operation is bigger; ResizeSmaller, when the area covered after the operation is smaller; and ResizeShape when the area covered is the same but the values of the properties are differents.

Move. Consists in changing the position of a component. It modifies the $PosX$ and $PosY$ properties. When $(PosX_i \neq PosX_{i+1}) \wedge (PosY_i = PosY_{i+1})$ the component has been displaced horizontally, when $(PosX_i = PosX_{i+1}) \wedge (PosY_i \neq PosY_{i+1})$ the component has been displaced vertically and finally, when $(PosX_i \neq PosX_{i+1}) \wedge (PosY_i \neq PosY_{i+1})$ the component has been displaced both horizontally and vertically.

3   Database Design for Storing Interactions

When an interaction occurs in the mashup application, a data acquisition process will start and save all the information regarding the operation by the interaction. Together with the operation it is convenient to save the information about the user that generates the interaction as well as the component that is affected. It is also advisable to save the state that remains in the workspace after the operation. Although storing all the workspace might seem rather costly, it would make it possible to rebuild all the users' behavior step by step throughout the interfaces in case further analysis, not considered at design time, is required. That is why this option is viewed in this proposal.

In order to store data of the interactions that have been performed by users in the mashup UI, it is necessary to define a relational database model that would be able to store all the relevant information of the interaction. This relational database should be as complete as possible to have a good understanding of the interaction itself and the circumstances that surround that interaction. Figure 2 shows a proposal relational database schema that represents the interaction performed as well as the situation of the UI after it.

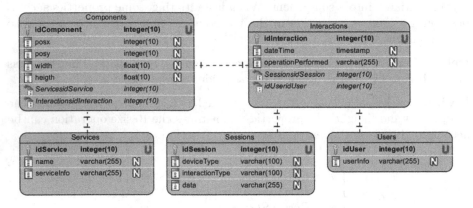

Fig. 2. Database schema to storage interaction in a standard mashup UI

Each row of the *Interactions* table corresponds to an interaction taken by the user and the field *operationPerformed* saves the kind of operation performed. The *Interactions* table is related to the *Sessions* table, thus all the operations performed in the same session are grouped. The *Sessions* table has two important fields *deviceType* and *interactionType*. The first one storages the kind of device used in the session where the operations are performed; it can be a Tablet, a Laptop, a SmartPhone, Home Automation Systems or Smart Watches, among others. The second one storages the type of interaction used when performing the operation: mouse, keyboard, gesture, voice or presence, among other. Note that there can be many sessions working currently because one application can be use at the same time through more than one UI from the same or different devices or systems.

The *Users* table, which is related to the *Interactions* one, has information of all users registered in the application. Usually, there are a lot of users registered in most of applications, therefore, it is important to distinguish the operation performed by each one of them. Some information systems allow guest users to access to the system and, for those kind of users, there is normally a specific row in the *Users* table. Note that it would be necessary to obtain extra information about users that do not come with the interaction, hence it would be useful a request to Web services provided by the service, if any. In case the mashup UI has no users registered the *Users* and *Sessions* tables can be thrown out.

The *Components* table includes all components that populate the workspace after and interaction has been performed. With the information gathered in this table, it is possible to rebuild the workspace exactly as it was when the interaction was performed. The *posx, posy, width* and *height* attributes are enough to set each component in the workspace. Finally, the *Services* table, related to the *Components* table, has information about all the services that are registered in the Information Systems. As in the *users* table, it could be necessary, but not mandatory, to access to external Web services to obtain more relevant information about services that do not come with the interaction.

This database schema could seem rather costly due to the extensive resources consumption that may involve to store the workspace with all the components contained within. However, it is optimized in the sense that the database is expressive enough not only to generate datasets with rich data for further analysis, but for recreate user interaction step by step in case that more information about any aspect of the interaction could be detected as needed to infer a knowledge not contemplated when designing the database schema.

4   Database Behavior in a Real Mashup

Once the relational database schema proposed has been deployed in a database over a real environment, it is possible to access to all the interactions that have occurred with a great detail. In this case, we deployed the relational database in ENIA, the mashup interface previously described, which focuses on the management of environmental information. The real implementation of the database has more tables and the mashup interface has more operations compared to the set of fields and operations we have discussed previously in this paper. But, as a matter of fact, the operations and tables described are present. The next piece of SQL code queries a MySQL database deployed in a platform as a service cloud infrastructure provided by Azure. ClearDB provides the MySQL databases in Azure as database as a service. This query extracts all the operations that have been performed by users and sessions specifying in each case the kind of UI upon which the interaction was performed (browser, mobile browser, tablet app, smartphone app...) as well as the type of interaction used to perform the operation (mouse, keyboard, touch, gesture, voice...).

```
SELECT interactions.idInteraction, interactions.dateTime,
    interactions.operationPerformed, interactions.Sessions_idSession,
    interactions.Users_idUser, sessions.deviceType,
    sessions.interactionType
FROM interactions, sessions, users
WHERE interactions.Users_idUser=users.idUserClient AND
    interactions.Sessions_idSession=sessions.idSession
```

Figure 3 presents the data obtained from the SQL code shown before. We can be distinguish between add, move and delete operations. All of them have been performed in a desktop or laptop browser and the form of interaction has been made by touching the laptop or computer screen.

idInteraction	dateTime	operationPerformed	Sessions_idSession	Users_idUser	deviceType	interactionType
9631	2016-03-08 11:23:57	Add	691	1	Browser	Touch
10271	2016-03-09 09:15:12	Add	711	1	Browser	Touch
9611	2016-03-07 13:39:55	Add	671	31	Browser	Touch
12961	2016-03-11 18:06:21	Add	681	31	Browser	Touch
12971	2016-03-11 18:06:21	Add	681	31	Browser	Touch
12981	2016-03-11 18:06:21	Add	681	31	Browser	Touch
13021	2016-03-11 18:08:49	Move	741	31	Browser	Touch
13031	2016-03-11 18:08:57	Move	741	31	Browser	Touch
13041	2016-03-11 18:10:09	Delete	751	31	Browser	Touch
13051	2016-03-11 18:21:42	Add	681	31	Browser	Touch
13061	2016-03-11 18:21:42	Add	681	31	Browser	Touch
13071	2016-03-11 18:21:43	Add	681	31	Browser	Touch
13081	2016-03-11 18:21:43	Add	681	31	Browser	Touch
13091	2016-03-11 18:21:43	Add	681	31	Browser	Touch
12101	2016-03-11 18:21:43	Add	681	31	Browser	Touch

Fig. 3. Results from the query to the interaction db

This database allows to access to every operation performed in the UI from heterogeneous devices; it also enables to recreate the user behavior step by step by analyzing the *workspace*, as it has been later each operation performed, for a better understanding of the user's behavior.

5   Conclusions and Future Work

This paper proposes a relational database to allow mashup UIs to store the interaction performed by users over them. The database is valid even when the interface runs in distributed heterogeneous devices that support different interactions modes. A clear definition of a mashup GUI morphology has been made in order to study it and suggest a relational database that can save the interaction with accuracy.

The creation of a data acquisition process is proposed as future work. This data acquisition process could be a microservice than runs in the cloud and is continuously listening to the request from different mashup UIs distributed in multiple devices. It can just receive all the data directly from the client and store it or even make some requests to the mashup UI services, if any, to obtain extra information about users or components. Moreover, it can make a request to third party services that can provide valuable context information.

The information stored in the database proposed contains valuable information about the users' behavior. Machine learning experiments can be performed for the creation of an automatic learning system that can be used to offer a better user experience. The discovery of behavioral patterns gives the opportunity to create prediction models that assist users, providing them with the components they are most likely to need, including the shape, size and layout configuration properties they expects.

A future deployment of the data acquisition process that storage the interaction in the relational database proposed and the machine learning experiments can be used over the work shown at Criado *et al.* [11], where component-based interfaces are adapted at run time using model transformation according to a set of rules. The new rules generated can update the rules repository making the application autonomously evolve over time [12].

Acknowledgments. This work was funded by the EU ERDF and the Spanish Ministry of Economy and Competitiveness (MINECO) under Project TIN2013-41579-R and under a FPI Grant BES-2014-067974 and by the Andalusian Regional Government (Spain) under Project P10-TIC-6114. Wang has been funded by the US National Science Foundation.

References

1. Elmqvist, N.: Distributed user interfaces: state of the art. Distributed User Interfaces DUI 2011. Human–Computer Interaction Series, pp. 1–12. Springer, London (2011)
2. Daniel, F., Matera, M.: Mashups - Concepts, Models and Architectures. Springer, Heidelberg (2014)
3. Vallecillos, J., Criado, J., Padilla, N., Iribarne, L.: A cloud service for COTS component based architectures. Comput. Stand. Interfaces **48**, 198–216 (2016). Elsevier
4. Fernández-García, A.J., Iribarne, L.: TDTrader: a methodology for the interoperability of DT-Web Services based on MHPCOTS software components, repositories and trading models. In: 2nd International Workshop of Ambient Assisted Living (IWAAL2010), pp. 83–88 (2010)
5. Roscher, D., Lehmann, G., Schwartze, V., Blumendorf, M., Albayrak, S.: Dynamic distribution and layouting of model-based user interfaces in smart environments. In: Hussmann, H., Meixner, G., Zuehlke, D. (eds.) Model-Driven Development of Advanced User Interfaces. SCI, vol. 340, pp. 171–197. Springer, Heidelberg (2011)
6. Hoyer, V., Fischer, M.: Market overview of enterprise mashup tools. In: Bouguettaya, A., Krueger, I., Margaria, T. (eds.) ICSOC 2008. LNCS, vol. 5364, pp. 708–721. Springer, Heidelberg (2008)
7. Geckoboard. Commercial mashup KPI dashboard. https://www.geckoboard.com/
8. Cyfe. Commercial mashup business dashboard. https://www.cyfe.com/
9. The Andalusian Environmental Information Network (REDIAM). http://www.juntadeandalucia.es/medioambiente/site/rediam
10. ENIA. Environmental Inf. Agent. http://acg.ual.es/projects/enia/ui/
11. Criado, J., Rodríguez-Gracia, D., Iribarne, L., Padilla, N.: Toward the adaptation of component-based architectures by model transformation: behind smart user interfaces. Softw.: Pract. Experience J. **45**(12), 1677–1718 (2015)
12. Fernandez-Garcia, A.J., Iribarne, L., Corral, A., Wang, J.Z.: Evolving mashup interfaces using a distributed machine learning and model transformation methodology. In: Ciuciu, I., et al. (eds.) OTM 2015 Workshops. LNCS, vol. 9416, pp. 401–410. Springer, Heidelberg (2015). doi:10.1007/978-3-319-26138-6_43

Virtual Spatially Aware Shared Displays

Felix Albertos-Marco$^{(\boxtimes)}$, Victor M.R. Penichet, and Jose A. Gallud

Computer System Department Albacete,
University of Castilla-La Mancha, Albacete, Spain
{felix.albertos,victor.penichet,jose.gallud}@uclm.es

Abstract. Nowadays, sharing resources over multiple devices is a common task. Some approaches consist in sharing a common workspace among users, or moving user interface elements between displays. But distributing interaction between displays is critical in cross-device environments. In this work, we present a technique for distributing content and devices in shared workspaces using cross-device displays. This technique, referred to as the virtual spatially aware technique, allows the creation of virtual shared displays and the coordination of cross-device interactions. By using this technique, we propose a method for arranging content and devices on virtual displays. We also present a prototype that supports the virtual spatially aware technique. This prototype has been built using web technologies, and it is able to run in any modern web browser.

Keywords: Distributed interaction · Virtual spatially aware · Shared workspaces · Content representation

1 Introduction

There are many approaches for the distribution of interaction among multiple devices. Albertos et al. presented synchronous interaction over shared resources in Drag&Share [1]. They provided a shared workspace which is the same for all users. Other approaches are aimed at moving user interface elements between displays according to a device's characterization [4]. But these approaches do not take into account the arrangement of cross-device displays within a shared display.

Radle et al. [3] initiated the debate about the use of spatially-aware (use of real-world spatial configurations as the referential domain) or spatially-agnostic cross-device interaction techniques. Their results showed that spatially-aware techniques are preferred by users. There is also no consensus on how to make the mapping between input and output in applications using multi-display and/or cross-device interactions.

In this work we present a technique for the design of virtual spatially-aware cross-device interaction. Using a planar technique, which is easy to match to the space [2], we propose the use of virtual spatially aware shared displays for supporting the creation of shared interaction spaces using cross-device displays.

© Springer International Publishing AG 2016
S. Casteleyn et al. (Eds.): ICWE 2016 Workshops, LNCS 9881, pp. 196–199, 2016.
DOI: 10.1007/978-3-319-46963-8_19

This new technique is based on the arrangement of devices and resources using a virtual display and distributing interaction over multiple devices. It is intended to be used to share resource interaction on multiple displays or to facilitate the set-up of multi-display screens.

2   Arranging Content on Virtual Spatially Aware Shared Displays

The use of virtual spatially aware shared displays allows the management of devices and resources in a common interaction area. This is achieved through the use of virtual elements that represent devices and resources that are virtually arranged on the virtual display. This arrangement might depend on physical or logical conditions.

Figure 1 shows a virtual display on which multiple resources are shared among multiple devices, which are represented by a semi-transparent gray square with a number representing their ID. The device will display the resources according to the representation of the virtual display. For example, the device with ID 122 will show the picture of a graph. These representations can be moved freely within the virtual display to show other resources or to make compositions with other displays. For example, real displays which are represented by IDs 102, 107, 112 and 117 are arranged as one big display. Therefore, they will show the area of the virtual display as a real and larger display, with the corresponding resources on it.

A support tool has been developed to support the virtual spatially aware shared displays. This tool allows the management of virtual displays, devices and content. In the following figures the virtual display is shown within a green square, device displays are in a pink square, and resources are shown within a blue square. Mapping between virtual and real elements is represented using arrows.

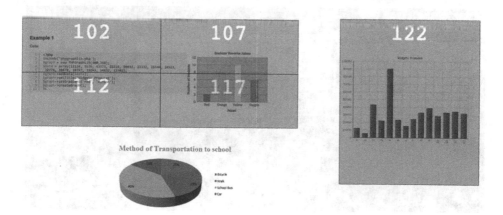

Fig. 1. Virtual display sharing resources among multiple devices

Figure 2 shows a scenario with a computer managing the virtual display and other devices (laptop, desktop pc and mobile phone) connected to the virtual display. It is worth pointing out that on the computer that manages the virtual display there are two browsers representing devices connected to the virtual display. There are no limitations on the location of virtual or device displays. Both systems work over web technologies and only require a modern browser to be run. Devices are connected through a browser to the virtual display.

Figure 3 shows how resources are managed using a virtual display. The set-up consists of two displays connected to two computers and another computer that hosts the virtual display. The resource in the virtual display is an image, but it can be any web resource. The devices on the virtual display are arranged so that the image is between them. As a result, the image generated by the two displays corresponds to the arrangement on the virtual display. The image is within a blue square, both on the real multi-display and in the virtual set-up. In addition, any interaction with the image on the virtual display or on the connected devices will be reflected across the entire environment (e.g. the movement of the image on any device).

Fig. 2. Mapping between devices and the virtual display

Fig. 3. Two displays arranged on the virtual display showing a resource

3   Conclusions and Future Work

In this work we have presented a tool for supporting the virtual spatially aware shared displays interaction technique. This technique allows the arrangement of content and devices on a virtual display. It supports cross-device interactions. As future work, we will continue with the development of the support tool. This development will allow us to perform an analysis of more sophisticated interaction techniques within the system and also, among others, the study of the impact of positioning devices on the virtual display according to their real position in the physical world.

Acknowledgments. This work has been partially supported by grant 2014/10340 from the University of Castilla-La Mancha.

References

1. Marco, A.F., Penichet, V., Gallud, J.A.: Collaborative e-learning through drag & share in synchronous shared workspaces. J. Univ. Comput. Sci. **19**(7), 894–911 (2013). http://dx.doi.org/10.3217/jucs-019-07-0894
2. Nacenta, M.A., Gutwin, C., Aliakseyeu, D., Subramanian, S.. There and back again: cross-display object movement in multi-display environments. J. Hum.-Comput. Interact. **24**(1), 170–229 (2009). http://pdfserve.informaworld.com/764323_770885140_910602221.pdf
3. Radle, R., Jetter, H., Schreiner, M., Lu, Z., Reiterer H., Rogers, Y.: Spatially-aware or Spatially-agnostic? Elicitation and evaluation of user-defined cross-device interactions. In: Proceedings of the 33rd Annual ACM Conference on Human Factors in Computing Systems (CHI 2015), pp. 3913–3922 (2015). ACM, New York. http://doi.acm.org/10.1145/2702123.2702287
4. Yang, J., Widgor, D.: Panelrama: enabling easy specification of cross-device web applications. In: Proceedings of the SIGCHI Conference on Human Factors in Computing Systems (CHI 2014), pp. 2783-2792. ACM, New York. http://doi.acm.org/10.1145/2556288.2557199

Flexible Distribution of Existing Web Interfaces: An Architecture Involving Developers and End-Users

Sergio Firmenich[1(✉)], Gabriela Bosetti[1], Gustavo Rossi[1], and Marco Winckler[2]

[1] LIFIA, Facultad de Informática, Universidad Nacional de La Plata and CONICET,
La Plata, Argentina
{sergio.firmenich,gabriela.bosetti,
gustavo}@lifia.info.unlp.edu.ar
[2] ICS-IRIT, University of Toulouse 3, Toulouse, France
winckler@irit.fr

Abstract. This paper presents a novel approach towards the opportunistic and lightweight distribution of existent Web User Interfaces. We describe an architecture that allows end-users to collect UI objects into a distributed UI-Component-oriented PIM, accessible from different user's devices. Once in the PIM, the collected UI components are wrapped with different DUI-based behaviours that may be triggered by the user, as PIM's objects plug-ins. We present an overview of the architecture; some default DUI-based behaviours are introduced and illustrated through examples. Besides, we show how the architecture supports the development of new DUI-based behaviours.

Keywords: Client-side adaptation · DUI · End-user development

1 Introduction

The distribution of user interfaces (DUI) has been a growing trend in the last ten years. Beyond the understanding about how DUI applications can improve user experience [12], several works for engineering DUI Web applications have emerged [10, 13]. Even more, approaches for more specific cross-device interaction support has been defined, such as the use of Kinect [11]. However, it is still hard to find popular Web sites or applications supporting DUI.

When applications do not offer features that users may need, experience has shown that the crowd react trying to satisfy these needs by itself. This is a very common practice in Web Browsing Augmentation, i.e. using tools (deployed such as Web Browser Extensions) to augment Web application capabilities. To cite one example, simple solutions for cross-device interaction such as "Slides – Presentation Remote"[1] has more than sixty thousand users, just offering a remote control for presentations in some well-known Web applications (Google Drive, SlideShare, Prezi, etc.). Examples like this clearly show that while ad-hoc developers may create this kind of experience, there are also users expecting them.

---

[1] https://chrome.google.com/webstore/detail/slides-presentation-remot/mhfdnafbhfglkcjgk-goopjoadaopcomi.

© Springer International Publishing AG 2016
S. Casteleyn et al. (Eds.): ICWE 2016 Workshops, LNCS 9881, pp. 200–207, 2016.
DOI: 10.1007/978-3-319-46963-8_20

We started this work by analyzing how to apply the lessons learned in Web Augmentation [3] into the field of DUI. Web Augmentation comprises those approaches that aim to adapt content and functionality of existing (usually third-party) Web applications, once these are loaded on the client. This technique has been used successfully in different domains, such as mash-ups [17] or end-user driven Web tuning [4]. Web Browser Augmentation is a perfect target for some DUI applications, such as supporting opportunistic remote control, layout distribution or UI migration.

In this paper, we propose to involve not only developers but also end users in the process of user interface distribution. The main idea is to provide developers with an easy way to implement DUI layers over existing Web pages, while end users may decide how to apply such layers on their preferred Web sites. This is achieved by managing a UI-Component-oriented PIM, where users may collect the target UI components that then can be manipulated from the applications defined by developers. The paper presents the overall architecture and the supporting tools through some case studies.

The paper is structured as follows. First, we present the related works in Sect. 2. Section 3 introduces our approach and the main components of the underlying architecture. The supporting tools are illustrated via case studies in Sect. 4. Finally, we discuss our contribution in Sect. 5, in conjunction with our future work.

2   Background and Related Works

Since the early years of Web Augmentation [2] (WA) and Mash-Ups applications [5], several approaches have emerged for adapting or integrating existing Web contents. Diverse communities of users, both developers and amateur end users with technical know-how, set up the basis and contributed in the creation of new repositories of augmentation artefacts, which improved the Web with extra features that original Web applications did not contemplate. Such is the case of Greasyfork, or Mozilla addons. Some End User Development [9] (EUD) works arose later in those fields, to empower users to solve their particular needs by themselves. Concerning DUI in conjunction with the WA and EUD fields, and as pointed out in [14], we can appreciate that it is not easy to provide DUI for existing and third party Web content, although some approaches have tackled isolated specific dimensions, specifically involving end-users. User-Driven DUI was previously defined in [15]; however, although this approach does not consider third-party existing Web sites as potential targets, other approaches do. For instance, [7] lets users to annotate some parts of exiting Web UI in order to migrate components under user demand. The main idea is to allow users to access a Web application from a desktop environment, and then to continue the interaction with it from a mobile device, migrating those annotated portions of the UI. Other similar approaches using proxies are Proxy-work [16] and WebSplitter [8]. And there are also approaches addressing flexible interface migration [1].

In this paper, we present an architecture to apply DUI layers over existing Web content. But in contrast with the existing works, we do not provide an end-user tool for performing specific DUI applications, but an architecture to enable developers from Web Augmentation communities to easily define new kind of DUI applications, by taking advantage of the underlying distributed UI-Component-oriented PIM. However, since we share the philosophy behind empowering end-users with specific tools, our approach lets them to instantiate a DUI application in their preferred Web sites.

Our approach may support several kinds of DUI dimensions, such as light interface migration, remote control, distributed layout, etc. We have designed a client-site visual tool that lets users to specify how to distribute the UI of any existing Web site.

3   Distributing Web UI Objects

The main idea behind our approach is to provide end users with a specialized PIM with extra WA and DUI capabilities. Instead of collecting and structuring personal data such as traditional PIMs, we propose a UI-component-oriented PIM. This allows users to collect arbitrary DOM elements from existing Web sites. A DOM element represents a particular component of the user interface. These UI components are collected into the PIM and materialized as UIObjects. Our approach rests on the idea that a UIObject may be used by the end-user for triggering different behaviours supporting DUI. We call this kind of actions DUI behaviour. For instance, the user may activate the DUI behaviour "Close on other devices…" for a UIObject. This action would hide that UIObject in any other device registered in the platform by the user, and it would be functional only in the current device. Furthermore, although users may use some predefined "DUI behaviours", developers may create new ones. We call them behaviours because these are performed individually for a UIObjects; however, if several UIObjects are collected for a specific Web site, and different behaviours are executed with each of them, then a more complex DUI experience (involving combinations of objects) could be defined. In this way, at the end, the addition of several DUI behaviours may abstract a specific kind of distributed use of a Web site. For instance, a simple distributed layout could be supported if, after collecting several UIObjects, the user executes the "Close on other views…" behaviour for different UIObjects running in different devices/monitors. Meanwhile, other combined use of these behaviours may be oriented to control the UI displayed on a desktop computer from a mobile device. Since it is not our purpose to foreseen every possible DUI behaviour we propose instead a flexible architecture based on this UI-object PIM. The architecture was designed with two premises in mind: (1) users should be able to collect UIObjects into the PIM and use a specific DUI behaviour with them, and (2) developers should be able to create new kinds of DUI behaviours that may be added to the collected UIObjects. If the user does not select a behaviour, this is set by default according to the type of DOM element that was collected.

The overall idea is that by triggering some behaviour, the user is enriching an existing Web site with DUI features. In order to do that, users are allowed to annotate some existing DOM elements as UIObjects, that will be collected into the PIM and then, presented in another view. In this section, we introduce the components and concepts of the approach and some aspects about the architecture. Then we show how end-users may create and use UIObjects.

3.1   Underlying Architecture

As shown in Fig. 1, our approach mostly works at client-side. The UIObject-PIM is a Web Browser extension that, once installed in different devices, it allows users to share a unique space of information among them. In order to maintain the UIObjects synchronized, we provide a synchronization server (accessible from our Web Browser extension) that allows an instance (UIObject) of a particular UI component to be synchronized, by acting as a mediator among all the contexts where such instance is running. The approach is no constrained to a centralized server, the browser extension may be configured to work with any specific deployment of that server.

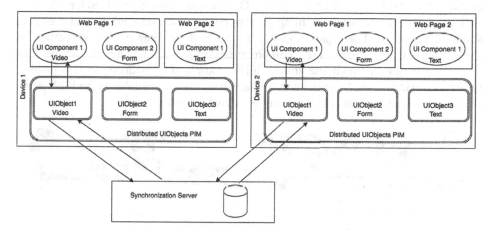

Fig. 1. Architecture schema of the UIObject-PIM platform

In this paper, we focus on the client-side features of this architecture, in which we can find three main artefacts:

- UI Component: in this way, we refer to the native UI components or widgets that compose a particular Web site. For instance, a UI Component could be a form, a video, a part of the DOM tree, etc. The main idea is that end-users may select under demand those UI Components of their interest. Through this selection, a UIObject is generated from a UI Component.
- UI Objects: these are representations of UI Components enriched with some behaviour that they do not have associated by default (in the original Web page). These UI Objects live in the UIObject-PIM.
- UI-Component-oriented PIM: this is a PIM oriented to maintain references to those UIObjects created by end users. This PIM requires authentication, so the user may collect and retrieve his preferred UIObjects from any of his devices. The UIObjects collected into the PIM are not just façades of UI Components, but managers of UIComponents with specific DUI-based behaviour added to its corresponding UIObject. This behaviour is materialized as operations that are executed for a particular UIObject, and it is executable directly by the user. For instance, if the user wants to render a UI Component only in one of his devices, then he must run the "Show only

in…" operation, that will ask the user which is the target device and then it will carry on the desired UI effect.

In this way, the PIM maintain the current state of the DUI model based on the operations (DUI-based behaviour) that were executed for the collected objects. In this way, the UI Component is like a view (in the same way as in the MVC pattern).

3.2   UIObjets and DUI-Based Behaviours in Detail

UIObjects are JavaScript objects abstracting UIComponents. During the abstraction process, the user may choose a target UIComponent. Figure 2 shows in (1), a user enabling the DOM selection and, in (2), a DOM element being highlighted for selection, with a proper context menu enabling the harvesting. Once selected, he can configure some of its aspects. For instance, give it a name and associate it some properties, as shown in (3). One of the most relevant properties is the UIComponentStereotype. This property allows users to choose among different kinds of UI widgets, such as images, text, forms, videos, etc. Once defined, all the UIObjects are available in the PIM, as the one in (4), so the user can interact with the offered behaviour.

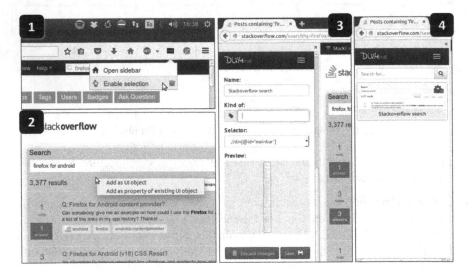

Fig. 2. Defining a UIObject

When a UIObject is collected into the PIM, the matching DUI-based behaviours (those preferred by the user) are attached to it under the basis of the decorator pattern [6]. There are two kinds of DUI-based behaviours. First, we can have stereotype-agnostic behaviours that may be attached to every UIObject since their goals are compatible with all kind of UIComponents. For instance, showing a particular UIObject only in a particular device. On the other hand, stereotype-specific behaviours are attached only to those UIObject that were collected as a specific UI stereotype, such as a YouTube Video. In this case, more specific operations such as *"Play video on…"* can be defined.

It is worth mentioning that the set of decorators applied to the UIObject can be configured by the end-user, by clicking the gear icon shown in Fig. 1; after doing this a context menu opens and the user has to select «manage» and finally «applied behaviour». The list of decorators is presented, and a series of parameters can be set.

Although we provide some basic behaviours, new ones may be defined by developers. To develop a new behaviour requires programming with client-side Web technologies (HTML, JavaScript, CSS), specially using a JavaScript API that allow developers to access, manage and add behaviour to UIObjects collected into PIM, considering that the communication mechanism among devices is solved transparently.

New behaviours may address new kind of DUI applications (for instance a particular kind of distributed layout), but may also perform specific operations wrapped in messages that the objects into the PIM may respond to. For instance, a developer may define a new behaviour for controlling a YouTube player by defining messages such as «*stop*» and «*play*». Then, if a user collects this object into the PIM and apply the decorator, then these messages could be sent from any device transparently. In these cases, developers are benefited with the underlying synchronization mechanism among PIMs state from different devices.

4   Case Study and Prototype

Although the approach lets developers to create new DUI behaviours, we have defined some predefined ones. So far we have identified: (1) opportunistic and volatile interface migration, (2) multi-monitor layout, (3) distributed layout, (4) messaged-based remote control, (5) navigation control and (6) remote UI Effect. For the sake of space, we are covering just one of them, in order to depict the feasibility of the approach.

At the time of this publication, we have partially implemented two prototype client-side tools supporting the approach. The first one is a Firefox 44 extension for desktop edition, the second one is a Firefox for Android extension, for version 42.0a1. At server-side, we implemented a minimal functionality to synchronize changes of the UIObjects both tools manage.

Consider a scenario where Máximo is living abroad for some months and he wants to write about every detail of his experiences in his blog. There are moments when he can write an entire entry from his computer, but he must migrate the task to his phone often, because his job requires constant mobility. His mother tongue is Spanish, but he is writing the blog in English. Sometimes, he have some doubts about language expressions, so he checks it out in linguee.com. It is desirable for him to have a mean for distributing some elements of the Linguee and Blogger Web sites.

Figure 3 shows our tool sidebar, with some UIObjects created from both sites; a search box and a results box from Linguee, and then a Blogger's entry main options, and a text area. So, when Máximo has to leave his house, he may open such elements in his mobile, the«XT1021» listed device, as shown in the left-bottom area of Fig. 3. Then, from his device, the browser extension will be notified to open the same URL and just open the defined UIObjects the user has defined for such URL.

Fig. 3. Distributing UI components from a web application to mobile

While the user changes the document, such modifications are notified to our server module, who asks every registered listener to update the view of the proper element. As we can see, the site functionality is not altered by our adaptation: the blog is still functional (you can see the automatic saving being executed in the right screen of Fig. 3). The advantages of accessing the blog through our platform is that: (1) Blogger does not provide a mobile version, so the same interface elements are presented in small devices; (2) While Blogger offers the ability to save changes in the entry, it does not offer the ability to keep the content up-to-date from two or more devices, and without our platform, it is required for every device to refresh the Web page every time they want to see the changes from another device.

5 Conclusions and Future Works

In this paper, we have presented an architecture for supporting the development of Web Browser Augmentation artefacts for the field of distributed user-interfaces. Browser augmentation oriented to DUI was also propose in previous works [7, 16]. We share the philosophy behind them, but we think that providing developers with an architecture that abstracts the more complicated technical issues underlying to DUI applications may help to create several kinds of new experiences by developing new DUI-based behaviours. Besides that, the same architecture contemplates a mechanism to allow end-users to instantiate those behaviours opportunistically in any Web site.

Currently, we are completing the implementation of the support system, which is partially operative. Once it is finished, we plan to carry on an evaluation with end-users in order to measure how useful and easy is for them to use our tools. We also plan to analyse new possible behaviours and study how this approach may raise new possibilities in the context of Web Augmentation, mash-ups and collaborative environments.

Acknowledgments. This work was supported by STIC AMSUD project WAMAW-OUR: Web Augmentation Methods for Adapting Web Sites for Supporting Opportunistic User Requirements

References

1. Bandelloni, R., Paternò, F.: Flexible interface migration. In: Proceedings of the 9th International Conference on Intelligent User Interfaces, pp. 148–155. ACM, January 2004
2. Bouvin, N.O.: Unifying strategies for web augmentation. In: Proceedings of the Tenth ACM Conference on Hypertext and Hypermedia: Returning to Our Diverse Roots: Returning to Our Diverse Roots, pp. 91–100. ACM, February 1999
3. Díaz, O., Arellano, C.: The augmented web: rationales, opportunities, and challenges on browser-side transcoding. ACM Trans. Web (TWEB) 9(2), 8 (2015)
4. Díaz, O., Arellano, C., Aldalur, I., Medina, H., Firmenich, S.: End-user browser-side modification of web pages. In: Benatallah, B., Bestavros, A., Manolopoulos, Y., Vakali, A., Zhang, Y. (eds.) WISE 2014, Part I. LNCS, vol. 8786, pp. 293–307. Springer, Heidelberg (2014)
5. Ennals, R.J., Garofalakis, M.N.: MashMaker: mashups for the masses. In: Proceedings of the 2007 ACM SIGMOD International Conference on Management of Data, pp. 1116–1118. ACM, June 2007
6. Gamma, E., Helm, R., Johnson, R., Vlissides, J.: Design Patterns: Elements of Reusable Object-Oriented Software. Pearson Education, Upper Saddle River (1994)
7. Ghiani, G., Paternò, F., Santoro, C.: On-demand cross-device interface components migration. In: Proceedings of the 12th International Conference on Human Computer Interaction with Mobile Devices and Services, pp. 299–308. ACM, September 2010
8. Han, R., Perret, V., Naghshineh, M.: WebSplitter: a unified XML framework for multi-device collaborative web browsing. In: Proceedings of the 2000 ACM Conference on Computer Supported Cooperative Work, pp. 221–230. ACM, December 2000
9. Lieberman, H., Paternò, F., Klann, M., Wulf, V.: End-User Development: An Emerging Paradigm, pp. 1–8. Springer, Dordrecht (2006)
10. Melchior, J., Vanderdonckt, J., Van Roy, P.: A model-based approach for distributed user interfaces. In: Proceedings of the 3rd ACM SIGCHI Symposium on Engineering Interactive Computing Systems, pp. 11–20. ACM, June 2011
11. Nebeling, M., Teunissen, E., Husmann, M., Norrie, M.C.: XDKinect: development framework for cross-device interaction using kinect. In: Proceedings of the 2014 ACM SIGCHI Symposium on Engineering Interactive Computing Systems, pp. 65–74. ACM, June 2014
12. Santosa, S., Wigdor, D.: A field study of multi-device workflows in distributed workspaces. In: Proceedings of the 2013 ACM International Joint Conference on Pervasive and Ubiquitous Computing, pp. 63–72. ACM, September 2013
13. Schreiner, M., Rädle, R., Jetter, H.C., Reiterer, H.: Connichiwa: a framework for cross-device web applications. In: Proceedings of the 33rd ACM Conference Extended Abstracts on Human Factors in Computing Systems, pp. 2163–2168. ACM, April 2015
14. Vanderdonckt, J.: Distributed user interfaces: how to distribute user interface elements across users, platforms, and environments. In: Proceedings of XI Interacción, vol. 20 (2010)
15. Vandervelpen, C., Vanderhulst, G., Luyten, K., Coninx, K.: Light-weight distributed web interfaces: preparing the web for heterogeneous environments. In: Lowe, D.G., Gaedke, M. (eds.) ICWE 2005. LNCS, vol. 3579, pp. 197–202. Springer, Heidelberg (2005)
16. Villanueva, P.G., Tesoriero, R., Gallud, J.A.: Proxywork: distributing user interface components of web applications. In: DUI@ EICS, pp. 58–61, June 2013
17. Wong, J., Hong, J.I.: Making mashups with marmite: towards end-user programming for the web. In: Proceedings of the SIGCHI Conference on Human Factors in Computing Systems, pp. 1435–1444. ACM, April 2007

Author Index

Printed in the United States
By Bookmasters